No. 1402
$14.95

NETWORK SYNTHESIS

BY CHARLES A. VERGERS

TAB BOOKS Inc.
BLUE RIDGE SUMMIT, PA. 17214

FIRST EDITION

FIRST PRINTING

Library of Congress Cataloging in Publication Data

Vergers, Charles A.
   Network synthesis.

   Includes index.

1. Electric network synthesis.   I. Title.
TK454.2.V46   621.3815′3   81-18294
ISBN 0-8306-0085-X   AACR2
ISBN 0-8306-1402-8 (pbk.)

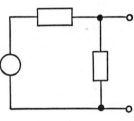

Contents

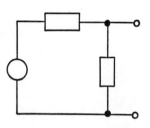

Introduction

The subject of analysis is an everyday term to the student of electronics. However, the term synthesis is normally more mysterious. To help take the fear out of network synthesis, I decided to write this book. The approach taken is the one I have used at the Capitol Institute of Technology in Kensington, Maryland, where I have taught network analysis and synthesis many times.

This book has many examples and illustrations in each chapter to help reinforce the information presented. It begins with a comparison of network analysis and network synthesis and then discusses the basic problem of recognizing transfer functions from their equation as well as from their frequency response. Material on deriving transfer functions of various networks is included. Once you have mastered this you will learn about the procedure for impedance and frequency scaling, Butterworth and Chebyshev low-pass filter derivation, the procedure for evaluating complex impedance and admittance equations in Chapter 6, and then about active network synthesis.

Various experiments that can be performed with a minimum amount of equipment are also included.

Although a good math background is helpful in understanding the material, advanced mathematical ideas are explained where needed. This book is suitable for students in a bachelor degree program in electrical engineering technology or electrical engineering.

I would like to thank Irene Shaffer of Hagerstown, Maryland for typing the manuscript for this book.

Other TAB books by the author:

No. 1132   *Handbook of Electrical Noise: Measurement & Technology.*

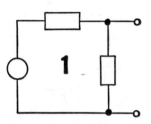

Network Analysis Theorems

In network analysis, the following problems may exist: (1) Given a network and the input signal. Find the output signal or some intermediate signal in the network; or, (2) Given a network and the output signal. Find the input or some intermediate signal in the network.

In network synthesis, the problem reduces to one of the following: (1) Given the input and output signal. Find the network; or, (2) Given the plot of the network frequency response. Find the network; or, (3) Given the network equation, which may be a transfer function with the form of the following:

a. Magnitude of the voltage or current transfer from input to output versus frequency.

b. Impedance versus frequency.

c. Admittance versus frequency.

From these plots, you must find the associated circuit.

In synthesis, there may be many different answers to the problem whereas in analysis there is only one.

There are various powerful theorems that allow us to simplify difficult problems in network analysis, namely, the voltage-divider theorem, the current-divider theorem, Thevenin's theorem, Norton's theorem, the superposition theorem, the reciprocity, the compensation, and the power transfer theorems.

VOLTAGE-DIVIDER THEOREM

This theorem allows us to determine the voltage across a component in a network by setting up a ratio of impedances (Fig. 1-1). The supply voltage is E_i. This voltage is connected across the series combination of R_1 and R_2. By simple Ohm's law we can state:

$$I = \frac{E_i}{R_1 + R_2}$$

Equation 1-1

Now we can ask, "What is the voltage across R_2?" Well, we know that the output is simply that given as

$$E_o = I \, (R_2)$$

Equation 1-2

By substitution of equation 1-1 into equation 1-2 we obtain

$$\frac{E_o}{R_2} = \frac{E_i}{R_1 + R_2}$$

Equation 1-3

This can be rewritten as

$$E_o = E_i \, \frac{R_2}{R_1 + R_2}$$

Equation 1-4

The voltage across the resistor R_2 is the product of the input voltage E_i and the ratio of R_2 to the sum of R_1 and R_2. It would not be necessary to calculate the current to find E_o, as can be seen from Equation 1-4.

Example 1-1. Determine the output voltage for the circuit in Fig. 1-2.

The resistance across which the output is developed consists of R_y in parallel with R_z. Since R_y and R_z are both 5 ohms, their

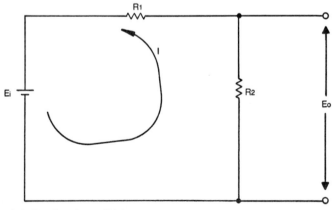

Fig. 1-1. Simple voltage divider.

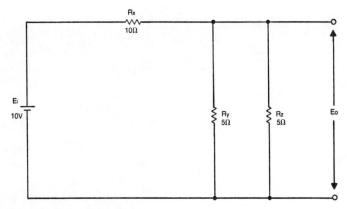

Fig. 1-2. Circuit for example 1-1.

parallel combination equals 2.5 ohms. The output voltage will be given by

$$E_o = E_i \frac{\dfrac{R_y \times R_z}{R_y + R_z}}{R_x + \dfrac{R_y \times R_z}{R_y + R_z}} = 10V \frac{2.5}{10 + 2.5}$$

$$= 10 \frac{2.5}{12.5} = 2V$$

Example 1-2. Determine the output voltage for the circuit in Fig. 1-3.

This circuit is slightly more complicated than the circuit in example 1-1, and we must deal with impedances, represented by symbol Z. Impedances are complex elements, consisting of both real (resistance) and imaginary (reactance) qualities. Impedance (Z) is commonly expressed as $Z = R + jX$, where R is resistance, X the reactance, and j is the rotational operator used to indicate inductive reactance $(+j)$, or capacitive reactance $(-j)$.

$$E_o = E_i \frac{Z_2}{Z_1 + Z_2} = 10V \frac{j50}{-j30 + j50} = 10V \frac{50}{20} = 25V$$

Notice that the output was larger than the input. This occurred with a passive network made of inductance and capacitance. Thus the output has been the result of a voltage developed across impedances.

Example 1-3. Determine the output voltage for the simple inductive network shown in Fig. 1-4:

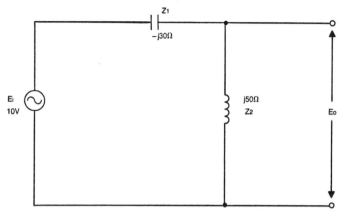

Fig. 1-3. Circuit for example 1-2.

$$E_o = E_i \frac{Z_2}{Z_1 + Z_2} = 100 \frac{j\omega L_2}{j\omega L_1 + j\omega L_2} = 100 \frac{L_2}{L_1 + L_2}$$

$$= 100 \frac{3}{4 + 3} = 100 \left(\frac{3}{7}\right) = 42.8 \text{ volts}$$

Example 1-4. Determine the output voltage for the simple capacitive network shown in Fig. 1-5:

$$E_o = E_i \frac{Z_2}{Z_1 + Z_2} = 100 \frac{\dfrac{1}{j\omega C_2}}{\dfrac{1}{j\omega C_1} + \dfrac{1}{j\omega C_2}}$$

$$= 100 \frac{1}{\dfrac{C_2}{C_1} + 1} = 100 \frac{C_1}{C_2 + C_1}$$

$$= 100 \frac{7\,\mu F}{3\,\mu F + 7\,\mu F} = 100 \left(\frac{7}{10}\right) = 70 \text{ volts.}$$

Notice that the formula has C_1 in the numerator while the denominator is just $C_1 + C_2$, therefore, we can say that the output voltage in a simple capacitive voltage divider (composed of two capacitors in series) is the product of the input voltage and the ratio of the capacitor (value) which does not have the output voltage across it to the sum of the two values. Notice that the inductive voltage divider and resistive voltage divider are essentially the

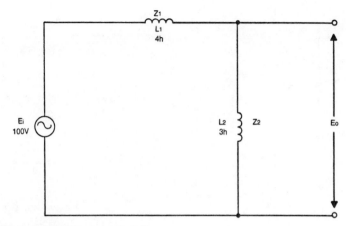

Fig. 1-4. Circuit for example 1-3.

same with regards to their dividing-ratio formula, while the capacitive voltage divider differs from them.

CURRENT-DIVIDER THEOREM

This theorem allows us to determine the current through a component by setting up a ratio between the components and multiplying by one of the circuit currents.

Refer to Fig. 1-6. Since we have a parallel circuit the voltage is the same across each component. Therefore we can state:

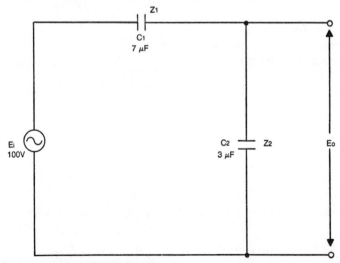

Fig. 1-5. Circuit for example 1-4.

5

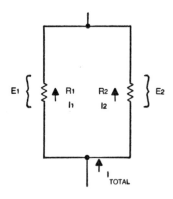

Fig. 1-6. Simple current divider.

$$E_1 = E_2 \qquad\qquad \text{Equation 1-5}$$

$$I_1R_1 = I_2R_2$$

We can then state an equation for I as:

$$I_1 = \frac{I_2R_2}{R_1} \qquad\qquad \text{Equation 1-6}$$

This is to be used in the case where we know R_1, R_2, and I_2. Likewise, we can write:

$$I_2 = \frac{I_1R_1}{R_2} \qquad\qquad \text{Equation 1-7}$$

This is to be used in the situation when R_1, R_2, and I_1 are known. Suppose we know the total current flowing into the circuit. The current in R_1 is I_1, and the current in R_2 is I_2. We can write an equation for the total current:

$$I_T = I_1 + I_2 \qquad\qquad \text{Equation 1-8}$$

Solving equation 1-8 for I_1, we have

$$I_1 = I_T - I_2 \qquad\qquad \text{Equation 1-9}$$

and solving for I_2 we have

$$I_2 = I_T - I_1 \qquad\qquad \text{Equation 1-10}$$

Now if we use these formulas in conjunction with the current-divider formula, we obtain equations 1-11 and 1-12 respectively:

$$I_T - I_2 = \frac{I_2 R_2}{R_1} \qquad\qquad \text{Equation 1-11}$$

6

$$I_T - I_1 = \frac{I_1 R_1}{R_2} \qquad \text{Equation 1-12}$$

These formulas may be rearranged to give:

$$I_T = \frac{I_2 R_2}{R_1} + I_2 \qquad \text{Equation 1-13}$$

$$= \left[\frac{R_2}{R_1} + 1 \right] I_2$$

$$I_T = \frac{I_1 R_1}{R_2} + I_1 = \left[\frac{R_1}{R_2} + 1 \right] I_1 \qquad \text{Equation 1-14}$$

Example 1-5. Calculate the current in branch one of the circuit in Fig. 1-7:

$$I_1 = I_2 \; \frac{R_2}{R_1} = 3A \; \frac{4k}{12k} = 1A$$

Example 1-6. Calculate the current in branch two of the circuit in Fig. 1-8:

$$I_2 = I_1 \; \frac{Z_1}{Z_2} \qquad |I_2| = 2mA \; \left| \frac{j60}{-j20} \right| = 6mA$$

THEVENIN'S THEOREM

This theorem is normally used when a voltage-source-equivalent circuit is required. This theorem permits a reduction of any complex two-terminal network to a simple network. To define, we may state the following: Any two-terminal linear network con-

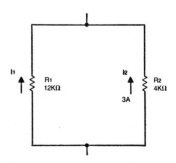

Fig. 1-7. Circuit for example 1-5.

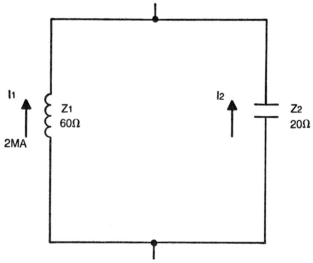

Fig. 1-8. Circuit for example 1-6.

taining energy sources and impedances can be replaced by an equivalent circuit composed of a voltage source in series with an impedance.

The procedure is three fold; refer to Fig. 1-9.

1. Find the voltage across the output terminals (points x-y). Using Ohm's law we have

$$E_{x-y} = E_{Th} = E_i \frac{Z_b}{Z_a + Z_b} \qquad \text{Equation 1-15}$$

2. Find the total impedance between points x and y, when the generator is shorted out.

$$Z_{x-y} = Z_{Th} = \frac{Z_a \times Z_b}{Z_a + Z_b} \qquad \text{Equation 1-16}$$

3. We then place the Thevenin's voltage, E_{Th}, and the Thevenin impedance, Z_{Th}, in the circuit of Fig. 1-10.

Example 1-7. Determine the Thevenin equivalent circuit for the network shown in Fig. 1-11.

$$E_{Th} = 100 \frac{40}{40 + 60} = 40 \text{ volts}$$

$$Z_{Th} = \frac{Z_a \times Z_b}{Z_a + Z_b} = \frac{(60) \times (40)}{60 + 40} = 24 \text{ ohms}$$

The resulting network is as shown in Fig. 1-12.

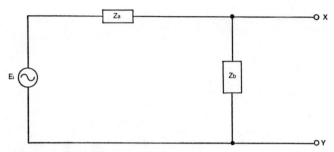

Fig. 1-9. Circuit to illustrate Thevenin's Theorem.

Example 1-8. Determine the Thevenin's equivalent circuit for the network shown in Fig. 1-13:

$$E_{Th} = E \frac{Z_b}{Z_a + Z_b} = 100 \frac{50}{j50 + 50}$$

$$E_{Th} = \frac{5000}{70.7 \underline{/45°}} = 70.7 \underline{/-45°} \text{ volts}$$

$$Z_{Th} = \frac{Z_a \times Z_b}{Z_a + Z_b} = \frac{j50 (50)}{j50 + 50} = \frac{j2500}{70.7 \underline{/45°}}$$

$$= \frac{2500 \underline{/90°}}{70.7 \underline{/45°}} = 35.3 \underline{/45°} \Omega$$

The resulting equivalent circuit is shown in Fig. 1-14.

Well, you may feel that this is a fine thing to know, but when would you use it in a practical problem? Refer to Fig. 1-15. Here you

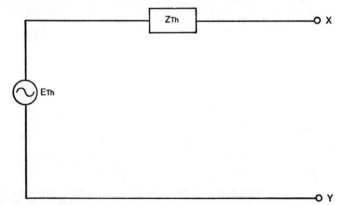

Fig. 1-10. Thevenin's circuit.

9

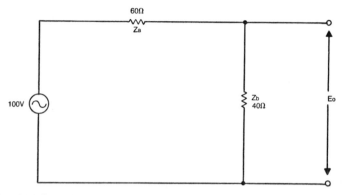

Fig. 1-11 Circuit for example 1-7.

see a circuit composed of a battery and three resistors. One resistor is in series with the other two which are in parallel.

Suppose you need to calculate the current in resistor R_3 for one hundred different values of R_3. There are two different ways you could do this.

First, determine the total current (I_T) in the circuit. For example, if $R_3 = 10$ ohms, you would have:

$$I_{TOTAL} = \frac{E_i}{R_T} = \frac{E_i}{R_1 + \frac{(R_2)\,(R_3)}{R_2 + R_3}} = \frac{100}{5 + \frac{(5)\,(10)}{5 + 10}}$$

$$= \frac{100}{5 + \frac{50}{15}} = \frac{100}{5 + 3.33}$$

$$I_{TOTAL} = 12 \text{ amperes}$$

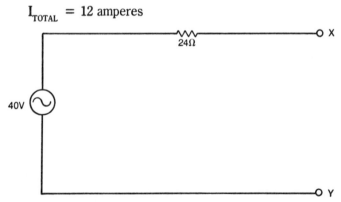

Fig. 1-12. Thevenin's circuit for example 1-7.

10

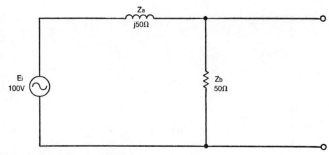

Fig. 1-13. Circuit for example 1-8.

Next, multiply the resistance of R_2 and R_3 in parallel, by the current I. This gives an output voltage of

$$E_o = I R$$
$$= (12)\ (3.33)$$
$$= 40 \text{ volts}$$

The current through resistor R_3 is then given by the following formula:

$$I_3 = \frac{E_o}{R_3} = \frac{40}{10} = 4 \text{ amperes}$$

Now you must repeat this for all of the remaining 99 values of R_3. This would take quite a long time unless you had a calculator programmed to perform this operation.

The other method makes use of Thevenin's Theorem.

1. Remove the resistor R_3. This gives the circuit as shown in Fig. 1-16. Compute E_{Th}:

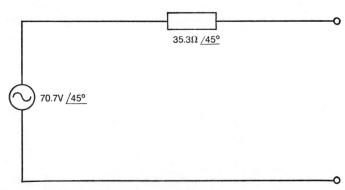

Fig. 1-14. Thevenin's circuit for example 1-8.

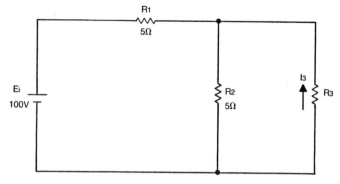

Fig. 1-15. Circuit for illustrating usefulness of Thevenin's Theorem.

$$E_{Th} = E \; \frac{R_2}{R_1 + R_2} = 100 \; \frac{5}{5 + 5} = 50 \text{ volts}$$

2. Compute R_{Th}:

$$R_{Th} = \frac{R_1 \, R_2}{R_1 + R_2} = \frac{5 \times 5}{5 + 5} = 2.5 \text{ ohms}$$

3. Place E_{Th} and R_{Th} in series giving the Thevenin's equivalent circuit as shown in Fig. 1-17.

4. Place the load on the circuit and calculate I_3.

$$I_3 = \frac{E_{Th}}{R_{Th} + R_3} = \frac{50}{2.5 + 10} = \frac{50}{12.5} = 4 \text{ amperes}$$

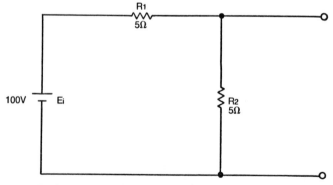

Fig. 1-16. Result of removing R3 in Fig. 1-15.

Now, for the remaining 99 resistors, all you need is this equation:

$$I_3 = \frac{50}{2.5 + R_3}$$

Thus, you can see that Thevenin's Theorem has simplified calculations significantly. It would be a waste of time to proceed in the previous manner.

NORTON'S THEOREM

This theorem is used mainly to reduce a network to a current-source-equivalent circuit. Interchanging current and voltage sources by utilizing this theorem and Thevenin's Theorem provides a powerful tool for network analysis.

As a definition, we could say that Norton's Theorem states that any two-terminal linear network containing energy sources and impedances can be replaced with an equivalent circuit consisting of a current source in parallel with an impedance.

The steps in determining Norton's equivalent circuit consists of the following:

1. Convert the circuit to Thevenin's equivalent circuit.

2. Find the effective short-circuit current, which is simply the Thevenin voltage divided by the Thevenin impedance.

3. Place the current generator in parallel with the impedance of the Thevenin's equivalent. This will cause an equivalent circuit which looks like that of Fig. 1-18.

Example 1-9. Determine Norton's equivalent circuit for the circuit shown in Fig. 1-19.

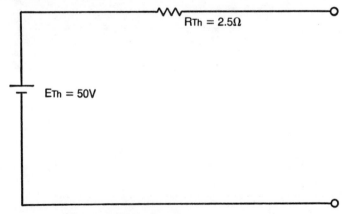

Fig. 1-17. Thevenin's circuit for Fig. 1-16.

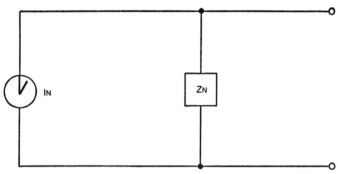

Fig. 1-18. Norton's Equivalent circuit.

1. Determine Thevenin's equivalent circuit, as was done in example 1-8.

2. Find the short-circuit current.

$$I_n = \frac{E_{Th}}{Z_{Th}} = \frac{50}{2.5} = 20 \text{ amperes}$$

3. Place the current generator in parallel with the 2.5-ohm impedance. The result is seen in Fig. 1-20.

Example 1-10. If a 10-ohm resistor was placed in parallel with the generator of Fig. 1-20, how much current would flow in the 10-ohm resistor? See Fig. 1-21.

Using the current divider theorem,

$$I_{TOTAL} = \left[\frac{R_2}{R_1} + 1 \right] I_2$$

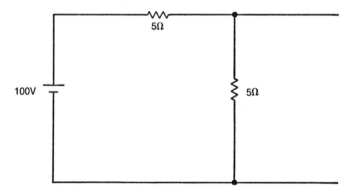

Fig. 1-19. Circuit for example 1-9.

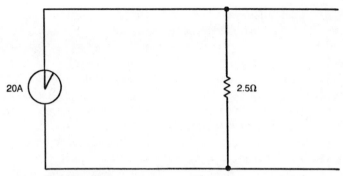

Fig. 1-20. Norton's equivalent circuit for example 1-9.

$$I_2 = \frac{I_T}{\frac{R_2}{R_1} + 1}$$

$$I_2 = \frac{2.0}{\frac{10}{2.5} + 1} = \frac{20}{4 + 1} = \frac{20}{5} = 4 \text{ amperes}$$

Notice that this is the same answer as obtained in the problem relating to Fig. 1-15.

SUPERPOSITION THEOREM

This theorem is used when several generators exist in a network. The generators may be of the same frequency or different frequencies. The theorem permits calculation of the total current in a network as the summation of the individual currents due to each generator signal. The circuit may be composed of both resistances and reactances, however, the reactances must be considered at the frequency of interest.

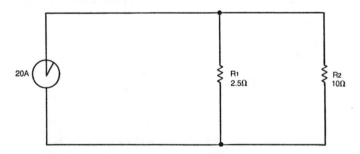

Fig. 1-21. Circuit for example 1-10.

15

As a definition, we could say that the Superposition Theorem states that the current that flows in any branch of a network of impedances, resulting from the simultaneous application of a number of electromotive forces which can be at any place in the network, is the algebraic sum of those component currents in that branch which would result from each source of voltage (emf) acting in an independent manner while all the other sources are replaced in the network by their respective internal impedances.

Example 1-11. Using the theorem of superposition, compute the current through resistor R_3 in Fig. 1-22.

1. Short out generator 2 and determine the voltage across R_3. See Fig. 1-23. Calculate Z_2 and R_3 in parallel:

$$Z_X = \frac{Z_2\,Z_3}{Z_2 + Z_3} = \frac{40\,(30)}{40 + 30} = \frac{1200}{70} = 17.1\Omega$$

Then:

$$E_{AB_1} = E_1\frac{Z_x}{Z_1 + Z_x} = 20\ \underline{/0^\circ}\ \frac{17.1}{j50 + 17.1}$$

$$= 20\underline{/0^\circ}\ \frac{(17.1)\,(-j50 + 17.1)}{(j50 + 17.1)\,(-j50 + 17.1)}$$

$$= 20\ \underline{/0^\circ}\ \frac{17.1\,(-j50 + 17.1)}{2500 + 293}$$

$$= \frac{342.8\,(-j50 + 17.1)}{2793} = -j6.1 + 2$$

This is the voltage across Z_3 due only to E_1.

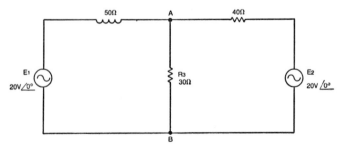

50Ω   A   40Ω

E1 20V$\underline{/0^\circ}$ R3 30Ω   E2 20V$\underline{/0^\circ}$

B

Fig. 1-22. Circuit for example 1-11.

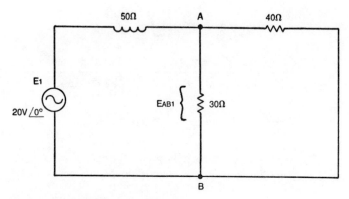

Fig. 1-23. Partial circuit for example 1-11.

2. Short out generator 1 and determine the voltage across Z_3. See Fig. 1-24. Calculate Z_1 and Z_3 in parallel:

$$Z_y = \frac{Z_1 Z_3}{Z_1 + Z_3} = \frac{j50 \,(30)}{j50 + 30}$$

$$E_{AB_2} = E_2 \frac{Z_y}{Z_y + Z_2} = 20 \underline{/0^\circ} \; \frac{\dfrac{(j50)\,(30)}{j50 + 30}}{\dfrac{(j50)\,(30)}{j50 + 30} + 40}$$

$$= 20 \underline{/0^\circ} \; \frac{(j50)\,(30)}{(j50)\,(30) + 40\,(j50 + 30)}$$

$$= \frac{20 \underline{/0^\circ} \; (j50)\,(30)}{j1500 + j2000 + 1200}$$

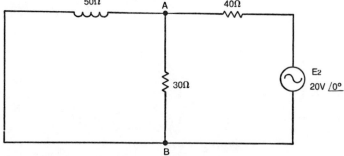

Fig. 1-24. Partial circuit for example 1-11.

$$= \frac{j30000}{j3500 + 1200} = \frac{j300}{j35 + 12}$$

$$= \frac{j300 \, (-j35 + 12)}{(j35 + 12) \, (-j35 + 12)} = \frac{-j10500 + 3600}{1225 + 144}$$

$$= \frac{-j10500 + 3600}{1369} = -j7.7 + 2.63$$

3. The total voltage across R_3 is simply the sum of the contributions of voltage due to E_1 and E_2 acting independently:

$$E_{Z3} = E_{AB1} + E_{AB2} = -j6.1 + 2 + -j7.7 + 2.63$$
$$= -j13.8 + 4.63 \text{ volts}$$

4. The current through Z_3 is then

$$I_3 = \frac{E_{Z3}}{Z_3} = \frac{-j \, 13.8 + 4.63}{30}$$

$$I_3 = -j0.46 + 0.154 \text{ amperes}$$

Example 1-12. Determine the total current in the circuit shown in Fig. 1-25. Note that the two ac generators have different frequencies. Thus you must individually compute the effects of each generator on the circuit.

Because the capacitor offers 40 ohms capacitive reactance at 60 Hz, it would offer half that value at 120 Hz. The impedances are as follows:

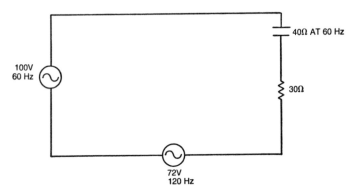

Fig. 1-25. Circuit for example 1-12.

$$Z_1 = j40 + 30 \text{ (for 60 Hz)}$$

$$Z_2 = j20 + 30 \text{ (for 120 Hz)}$$

We can determine the effective magnitudes as:

$$1Z_1 1 = \sqrt{(40)^2 + (30)^2} = 50 \text{ ohms}$$
$$1Z_2 1 = \sqrt{(20)^2 + (30)^2} \cong 36 \text{ ohms}$$

The current due to E_1 is:

$$|I_1| = \frac{1E_1 1}{1Z_1 1} = \frac{100}{50} = 2 \text{ amperes}$$

$$|I_2| = \frac{1E_2 1}{1Z_2 1} = \frac{72}{36} = 2 \text{ amperes}$$

Each current is due to a different generator and we cannot add their values directly. Instead, we must find the effective value of current. Each generator produces a certain amount of power dissipation in the load. The apparent power (P_a) due to E_1 is

$$P_{a_1} = I_1^2 Z_1$$

While the apparent power due to E_2 is

$$P_{a_2} = I_2^2 Z_2$$

In situations like this we compute the effective current squared which is

$$\frac{P_{a_1}}{Z_1} + \frac{P_{a_2}}{Z_2} = I_1^2 + I_2^2$$

Taking the square root, we get

$$I_{TOTAL} = \sqrt{I_1^2 + I_2^2} = \sqrt{(2)^2 + (2)^2} = 2.828 \text{ amperes}$$

MAXIMUM POWER TRANSFER THEOREM

This theorem states that maximum power is delivered to a load when the load is the conjugate of the generator impedance.

In other words, maximum power will be delivered to a load impedance if the load impedance is equal to the conjugate of the impedance looking back from the open-circuited load terminals of the network with all generators shorted and replaced by their respective internal impedances.

If you are not familiar with the term "conjugate," it simply is that number whose imaginary part is opposite in sign to that of the number in question. For example,

A	+ jB	is the conjugate of A-jB
4	+ j3	is the conjugate of 4-j3
−5	− j9	is the conjugate of −5+j9

Example 1-13. What value of Z_L (load impedance) will allow maximum power transfer in the circuit shown in Fig. 1-26. The circuit could be redrawn to that of Fig. 1-27.

If we remove the load, Z_L, we can use Thevenin's theorem and redraw the circuit as in Fig. 1-28.

If we remove the load, Z_L, we can use Thevenin's Theorem and redraw the circuit as in Fig. 1-28.

$$E_{Th_1} = E\left(\frac{Z_1}{Z_1 + Z_2}\right) = 30\left(\frac{20}{20 + 10}\right) = 20 \text{ volts}$$

$$E_{Th_2} = E\left(\frac{Z_3}{Z_3 + Z_4}\right) = 30\left(\frac{10}{10 + 20}\right) = 10 \text{ volts}$$

The effective Thevenin impedances are

$$Z_{Th_1} = \frac{Z_1 Z_2}{Z_1 + Z_2} = \frac{(20)(10)}{20 + 10} = 6.7 \text{ ohms}$$

$$Z_{Th_2} = \frac{Z_3 Z_4}{Z_3 + Z_4} = \frac{(10)(20)}{10 + 20} = 6.7 \text{ ohms}$$

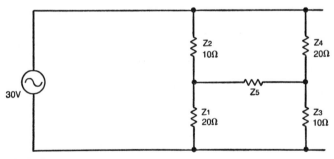

Fig. 1-26. Circuit for example 1-13.

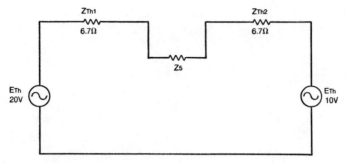

Fig. 1-27. Redrawn circuit for Fig. 1-26.

The Thevenin equivalent circuit can be drawn as in Fig. 1-29. The total generator resistance is then 13.4 ohms, and the generator voltage would be the difference between E_{Th_1} and E_{Th_2}, which is 10 volts.

For maximum power transfer, the load impedance must be the conjugate of the network impedance. However, all the impedances are real in this example. Therefore, the load should be equal to the network resistance for maximum power transfer, or 13.4 ohms.

You can now ask a different question: What is the maximum power that can be delivered to the load? Maximum power transfer occurs when the network impedance (includes generator impedance) and the load are conjugates. Thus, we could say

$$P_{MAX} = \frac{\left(E_{R_L}\right)^2}{R_L}$$

where E_{RL} would be 5 volts, since the generator impedance and load impedance are the same.

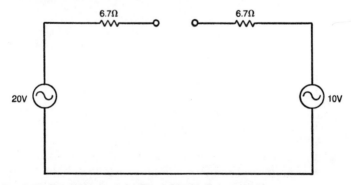

Fig. 1-28. Resulting circuit for Fig. 1-26 with Z5 removed.

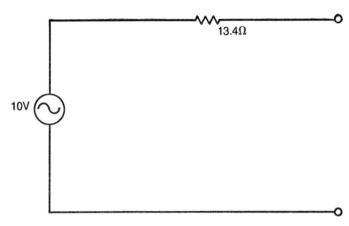

Fig. 1-29. Thevenin's circuit for example 1-13.

$$P_{MAX} = \frac{(5)^2}{13.4} = \frac{25}{13.4} = 1.865 \text{ watts.}$$

Example 1-14. What is the value of Z_L required for maximum power transfer in Fig. 1-30? The network impedance seen across the output terminals when the input is shorted is:

$$Z_0 = \frac{(4 + j3)(9 + j6)}{4 + j3 + 9 + j6} = \frac{36 + j27 + j24 - 18}{13 + j9}$$

$$= \frac{18 + j51}{13 + j9} = \frac{(18 + j51)(13 - j9)}{(13 + j9)(13 - j9)}$$

$$= \frac{234 - j162 + j663 + 459}{169 + 81} = \frac{693 + j501}{250}$$

$$= 2.77 + j2$$

The conjugate of the network impedance (Z^*) is the value of load impedance for maximum power transfer.

$$\begin{aligned} Z_L &= Z^* \\ &= (2.77 + j2)^* \\ &= 2.77 - j2 \text{ ohms.} \end{aligned}$$

RECIPROCITY THEOREM

This theorem allows the interchange of source and meter points for measurement purposes. Thus the source may be replaced when making an analysis on paper.

As a definition, we could state: If a voltage E is applied to any mesh of a network which is passive, linear, and bilateral, it produces a current I in another mesh of the same network. If the same voltage E is applied to the second mesh it will produce the identical current I in the first mesh.

To observe the use of this theorem examine the circuit in Fig. 1-31. We would like to know the current I_b flowing in impedance Z_L. We first must set up mesh equations.

$$E = I_a (Z_a + Z_b) - I_b (Z_b) \qquad \text{Equation 1-17}$$

$$O = (-I_a) (Z_b) + I_b (Z_b + Z_c + Z_L) \qquad \text{Equation 1-18}$$

The determinant of the system is

$$\Delta = \begin{vmatrix} Z_a + Z_b, & -Z_b \\ -Z_b, & Z_b + Z_c + Z_L \end{vmatrix} \qquad \text{Equation 1-19}$$

To find ΔI_b we write

$$\Delta I_b = \begin{vmatrix} Z_a + Z_b, & E \\ -Z_b, & 0 \end{vmatrix} \qquad \text{Equation 1-20}$$

To find I_b we write

$$I_b = \frac{\begin{vmatrix} Z_a + Z_b, & E \\ -Z_b, & 0 \end{vmatrix}}{\begin{vmatrix} Z_a + Z_b, & -Z_b \\ -Z_b, & Z_b + Z_c + Z_L \end{vmatrix}} \qquad \text{Equation 1-21}$$

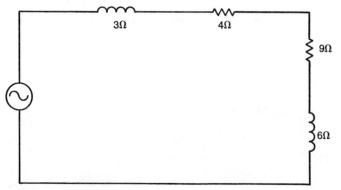

Fig. 1-30. Circuit for example 1-14.

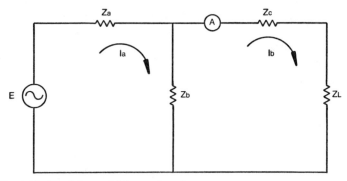

Fig. 1-31. Circuit for explaining reciprocity.

$$= \frac{(Z_a + Z_b)\,(0) - (-Z_b)\,E}{(Z_a + Z_b)\,(Z_b + Z_c + Z_L) - (-Z_b)^2}$$

$$= \frac{E\,(Z_b)}{(Z_a + Z_b)\,(Z_b + Z_c + Z_L) - Z_b^{\,2}}$$

$$= \frac{E\,(Z_b)}{Z_a Z_b + Z_a Z_c + Z_L Z_a + Z_b Z_c + Z_b Z_L}$$

The current I_b seems to be caused by an impedance of:

$$\frac{E}{I_b} = \frac{Z_a Z_b + Z_a Z_c + Z_L Z_a + Z_b Z_c + Z_b Z_L}{Z_b}$$

$$Z_x = Z_a + \frac{Z_a Z_c}{Z_b} + \frac{Z_L Z_a}{Z_b} + Z_c + Z_L$$

$$Z_x = Z_a + Z_c + Z_L + \frac{Z_a}{Z_b}\,[\,Z_c + Z_L\,]$$

Now, suppose we exchange the position of the ammeter and the voltage source E. Using Fig. 1-32, evaluate I_b.

$$E = I_a{}'\,(Z_b + Z_c + Z_L) - I_b{}'\,(Z_b)$$
$$0 = -I_a{}'\,(Z_b) + I_b{}'\,(Z_a + Z_b)$$

The determinant of the system is:

$$\Delta = \begin{vmatrix} Z_b + Z_c \times Z_L, & -Z_b \\ -Z_b, & Z_a + Z_b \end{vmatrix}$$

To find I_b' we write

$$\Delta I_b' = \begin{vmatrix} Z_b + Z_c + Z_L & E \\ -Z_b & 0 \end{vmatrix}$$

To find I_b' we write

$$I_b' = \frac{\begin{vmatrix} Z_b + Z_c + Z_L, & E \\ -Z_b, & 0 \end{vmatrix}}{\begin{vmatrix} Z_b + Z_c + Z_L, & -Z_b \\ -Z_b, & Z_a + Z_b \end{vmatrix}}$$

$$I_b' = \frac{(Z_b + Z_c + Z_L)\,0 - (-Z_b)\,(E)}{(Z_b + Z_c + Z_L)\,(Z_a + Z_b) - (-Z_b)^2}$$

$$= \frac{E\,(Z_b)}{(Z_a + Z_b)\,(Z_b + Z_c + Z_L) - Z_L{}^2}$$

$$= \frac{E\,(Z_b)}{Z_a Z_b + Z_a Z_c + Z_a Z_L + Z_b Z_c + Z_b Z_L}$$

Current I_b' seems to be caused by an impedance of

$$\frac{E}{I_b'} = \frac{Z_a Z_b + Z_a Z_c + Z_a Z_L + Z_b Z_c + Z_b Z_L}{Z_b}$$

$$Z_y = Z_a + \frac{Z_a Z_c}{Z_b} + \frac{Z_L Z_a}{Z_b} + Z_c + Z_L$$

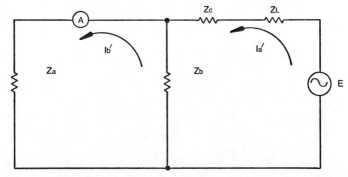

Fig. 1-32. Exchanging the ammeter and source in the circuit of Fig. 1-31.

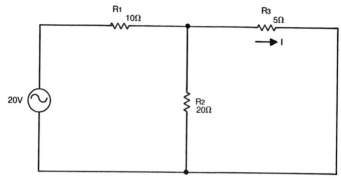

Fig. 1-33. Circuit for example 1-15.

$$Z_y = Z_a + Z_c + Z_L + \frac{Z_a}{Z_b}(Z_c + Z_L)$$

As we can see, Z_x and Z_y are the same. Since E was the same in each example, $I_b = I_b'$, the theorem of reciprocity is proved.

Example 1-15. Using the Fig. 1-33, determine the current I by using reciprocity techniques.

First, exchange the position of the generator and ammeter, as shown in Fig. 1-34. The resistance of the 10 and 20-ohm resistors in parallel is 6.66 ohms. The drop in voltage across the resistors in parallel is

$$E_x = E \frac{Rx}{Rx + R_3} = 20 \frac{6.66}{6.66 + 5} = 11.4 \text{ volts}$$

The current I would be

$$I = \frac{E_x}{R_1} = \frac{11.4}{10} = 1.14 \text{ amperes}$$

As a check, evaluate I by the conventional method. E and I are in the original solution (see Fig. 1-33).

The 20-1 and 5-ohm resistors are parallel. The resulting resistance is 4 ohms. The total current is

$$I_T = \frac{E}{R_T} = \frac{20}{10 + 4} = \frac{20}{14} = 1.43 \text{ A}$$

To determine the current I use the current-divider theorem:

26

$$I = I_T \left[\frac{R_2}{R_2 + R_3} \right] = 1.43 \, \frac{20}{25} = 1.14 \text{ amperes.}$$

This simple exercise illustrates the versatility of the reciprocity theorem.

In simple terms, we have evaluated the current in the 5-ohm resistor of Fig. 1-33 by moving the voltage source to the mesh that the 10-ohm resistor is in, and finding the current in the 10-ohm resistor.

COMPENSATION THEOREM

This theorem permits the determination of what will happen in a network if there is a variation of impedance in one branch of the network.

As a definition, we could say that the compensation theorem states that if a current I exists in a branch No. 1 of a network, and the impedance varies in that branch, the variation of voltage and current in any and every other branch of the network equals the voltage and current that would be caused by an opposing voltage equal to I multiplied by the impedance variation in the branch No. 1. This theorem is best explained with an example.

Example 1-16. A battery feeds power to two loads, namely, 4 and 6 ohms, through two resistors of 10 ohms each. Suppose a 2-ohm resistor is inserted into the 6-ohm branch so that it is in series with the 6-ohm resistor.

Find the change in current in branch X (see Fig. 1-35).

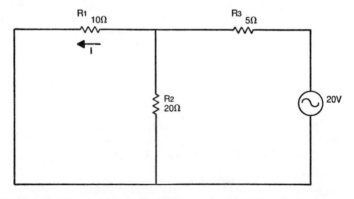

Fig. 1-34. Exchanging the ammeter and source in the circuit of Fig. 1-33.

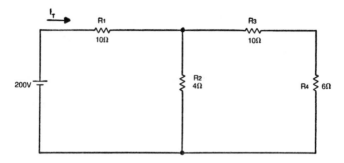

Fig. 1-35. Circuit for example 1-16.

$$I_T = \cfrac{E}{R_1 + \cfrac{R_2 \ (R_3 + R_4)}{R_2 + R_3 + R_4}} = \cfrac{200}{10 + \cfrac{4 \ (16)}{4 + 10 + 6}}$$

$$= \frac{200}{10 + 3.2} = \frac{200}{13.2} = 15.15 \text{ amperes.}$$

$$I_x = I_T \frac{R_2}{R_2 + R_3 + R_4} = 15.15 \ \frac{4}{4 + 10 + 6}$$

$$= 15.15 \ \frac{4}{20} = 3.03 \text{ amperes.}$$

If the 2-ohm resistor is added to the 6-ohm resistor, the incremental change in the voltage in branch X is

$$\begin{aligned} V &= I_x \ (Z) \\ &= (3.03) \times (2) \\ &= 6.06 \text{ volts} \end{aligned}$$

The impedance to which V is applied is:

$$Z = R4 + R3 + R + \frac{R_2 \ R_1}{R_1 + R_2} = 6 + 10 + 2 + \frac{4 \ (10)}{10 + 4}$$

$$= 18 + 2.86 \quad = 20.86 \text{ ohms}$$

The increase of current due to insertion of the extra resistance is

$$I = \frac{-V}{Z+} = \frac{-6.06}{20.86} = -.29 \text{ amperes.}$$

28

Because resistance was added, current must decrease. This is reflected in the negative sign.

The total current through branch X with the extra resistance is

$$I_{TOTAL} = I_x + I$$
$$= 3.03 - .29$$
$$= 2.74 \text{ amperes}$$

Let's check this answer another way. When the 2-ohm resistor is added, the circuit looks like that in Fig. 1-36. The total resistance would be

$$R_{TOTAL} = R_1 + \frac{R_2 (R_3 + R_4 + R)}{R_2 + R_3 + R_4 + R} = 10 + \frac{4 (10 + 6 + 2)}{4 + 10 + 6 + 2}$$

$$= 10 + \frac{4 (18)}{22} = 13.27 \text{ ohms}$$

The total current in the circuit is

$$I_{TOTAL} = \frac{E}{R_T} = \frac{200}{13.27} = 15.1 \text{ amperes.}$$

The current in branch X will be

$$I_x = I_T \frac{R_2}{R_2 + R_3 + R_4 + R} = 15.1 \frac{4}{4 + 10 + 6 + 2}$$

$$= 15.1 \frac{4}{32} = 2.74 \text{ amperes}$$

Therefore we see another proof of the problem.

Now let's work an analysis problem using the information we have studied so far.

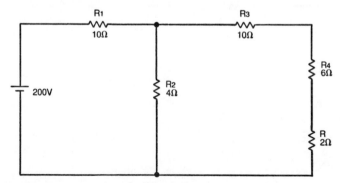

Fig. 1-36. Adjusted circuit for example 1-16.

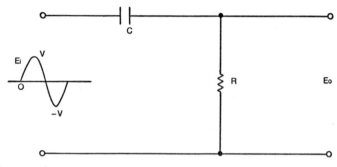

Fig. 1-37. Circuit for example 1-17.

Example 1-17. Determine the output signal for the network shown in Fig. 1-37. The input is a sinewave of the form

$$Vin = V \sin \omega t$$

If we use the voltage-divider theorem we obtain

$$\frac{E_o}{E_i} = \frac{R}{R + \dfrac{1}{j\omega C}}$$

This can be simplified to

$$\frac{E_o}{E_i} = \frac{j\omega CR}{j\omega CR + 1}$$

which can be placed in polar form

$$E_o = \frac{\omega CR \left/ 90 - \arctan\dfrac{\omega CR}{1}\right.}{\sqrt{(\omega CR)^2 + 1}}$$

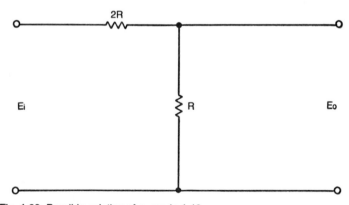

Fig. 1-38. Possible solution of example 1-18.

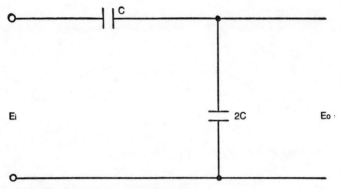

Fig. 1-39. Possible solution of example 1-18.

If we multiply the input voltage by the ratio $\dfrac{E_o}{E_i}$ we obtain the answer, which is:

$$E_o = (V \sin \omega t) \left(\frac{\omega CR \, \underline{/90 - \arctan \omega CR}}{\sqrt{(\omega CR)^2 + 1}} \right)$$

Now let us examine a simple synthesis problem.

Example 1-18. Suppose the desired input and output of a network are $E_i = V \sin \omega t$, and $E_o = \dfrac{V}{3} \sin \omega t$, respectfully. Find the network. The ratio of output to input is

$$H = \frac{E_o}{E_i} = \frac{\dfrac{V}{3} \sin \omega t}{V \sin \omega t} = \frac{1}{3}$$

Thus the network must have an output-to-input ratio of ⅓ with zero phase shift.

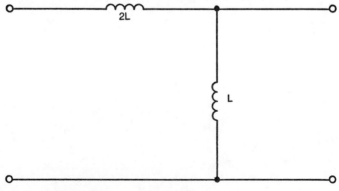

Fig. 1-40. Possible solution of example 1-18.

31

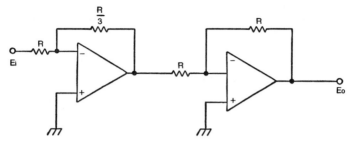

Fig. 1-41. Possible solution of example 1-18.

The possible solutions to this problem are too numerous to mention. Instead, I will give five solutions that could be used. Some of them are not very practical, but they would work. These are shown in Figs. 1-38 through 1-42. Every one of these networks has an output-to-input ratio of ⅓ with zero phase shift. The circuits of Figs. 1-38 through 1-40 are self explanatory. In Fig. 1-41, the first amplifier has voltage gain of

$$A_{V_1} = \frac{\dfrac{R}{3}}{R} = \frac{-1}{3}$$

The second amplifier has a gain of

$$A_{V_2} = \frac{-R}{R} = -1$$

The overall voltage gain is then

$$A_{V_T} = (A_{V_1}) = (A_{V_2}) = \frac{(-1)}{3} \ (-1) = \frac{1}{3}$$

In Fig. 1-42 the first amplifier has a voltage gain of

$$A_{V_1} = 1 + \frac{2R}{R} = 3$$

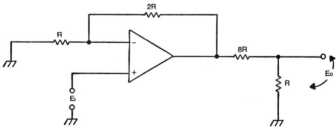

Fig. 1-42. Possible solution of example 1-18.

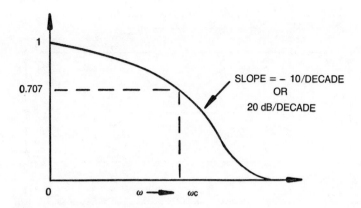

Fig. 1-43. Circuit for example 1-19.

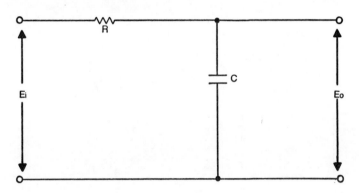

Fig. 1-44. Possible solution for example 1-19.

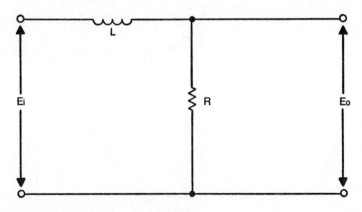

Fig. 1-45. Possible solution for example 1-19.

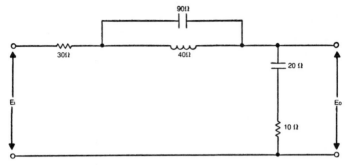

Fig. 1-46. Circuit for problem 1.

The gain of the resistive network connected to the output of the first amplifier is

$$A_{v_2} = \frac{R}{8R + R} = \frac{1}{9}$$

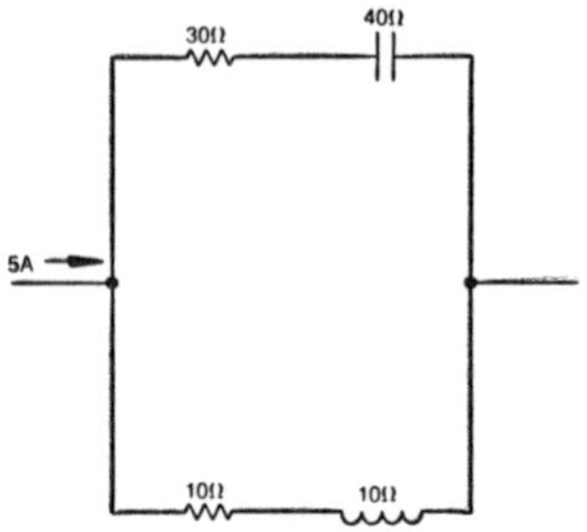

Fig. 1-47. Circuit for problem 2.

The gain of the overall circuit is

$$(A_{v_1})(A_{v_2}) = 3\frac{(1)}{9} = \frac{1}{3}$$

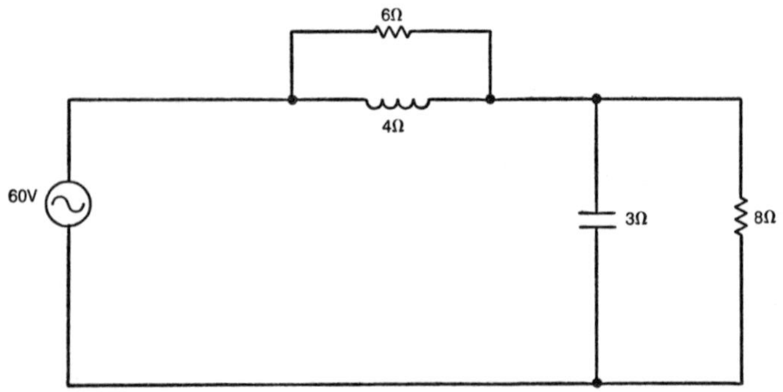

Fig. 1-48. Circuit for problem 3.

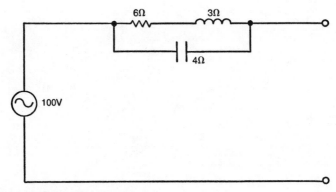

Fig. 1-49. Circuit for problem 4.

Example 1-19. Now let us look at another synthesis example. What network has a frequency response as shown in Fig. 1-43?

We can make the following observations:

1. Network has a response of one at low frequencies.

2. At the cutoff frequency the response is 0.707.

3. Beyond cutoff, the response drops by a factor of 10 for each decade, or 20dB per decade.

4. The network must be low-pass in nature. Some possible solutions are shown in Figs. 1-44 and 1-45. Each one of these networks has the same type of response.

This example required the use of the previous knowledge: You had to know the frequency-response behavior of low-pass networks.

CHAPTER SUMMARY

In this chapter, we reviewed some of the important theorems used in network analysis, and observed the differences between analysis and synthesis problems.

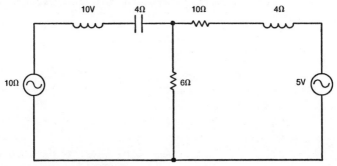

Fig. 1-50. Circuit for problem 5.

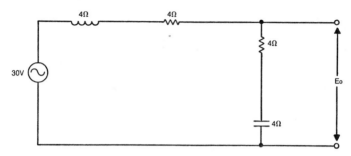

Fig. 1-51. Circuit for problem 6.

Problems

1. Using the voltage-divider theorem, compute the output voltage for the circuit of Fig. 1-46.

2. Find the current in every branch of the circuit of Fig. 1-47.

3. Determine the current in the 8-ohm resistor of Fig. 1-48 using Thevenin's Theorem.

4. Convert the circuit in Fig. 1-49 to Norton's equivalent.

5. Using the superposition theorem, find the current in the 6-ohm resistor of Fig. 1-50.

6. What is the value of load impedance which will allow for maximum power transfer in the circuit of Fig. 1-51.

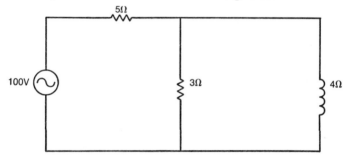

Fig. 1-52. Circuit for problem 7.

7. Given the circuit in Fig. 1-52, determine the current I in the inductor by using the reciprocity theorem.

8. A generator feeds power to two loads as shown in Fig. 1-35. If a 4-ohm resistor is inserted in branch three, find the change in current in that branch.

9. The input to a network is a sinewave: $V_{in} = 40 \sin \omega t$. The output voltage is given by: $V_o = 10 \sin \omega t$.

Determine at least four networks that would be solutions to the problem.

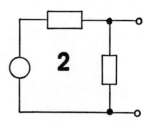

Basic Synthesis Problems

This chapter covers some basic synthesis problems.

BASIC TRANSFER FUNCTIONS

The transfer function of a network describes how the output behaves in respect to the input. This function has two parts, namely, **1.** magnitude and **2.** phase.

Although we did not call it a transfer function, when we examined the voltage-divider theorem we were really learning how to determine transfer functions. Actually a transfer function can represent the ratio of any two physical quantities. For example, suppose we had a machine in a lunch room that would give an apple whenever you put a dime in the machine. The transfer function would be:

$$H = \frac{OUTPUT}{INPUT}$$

Since one apple comes out for each dime that goes in the *magnitude* of the function would be one, while the *unit* of the function would be apple per dime. If you put in two dimes two apples would appear in the output. Therefore you could write output = (transfer function) (input). To find the number of apples in the output you multiply the transfer function by the number of dimes you put in the machine.

In network synthesis, the problem deals with more complicated transfer functions. Also, the units of the function may be ohms, mhos, or a dimensionless quantity.

One of the most important transfer functions deals with the ratio of output voltage to input voltage of a network.

Voltage Transfer Functions

All transfer functions have magnitude. To determine the magnitude of a function, find the size of the function as shown in the following example.

Example 2-1. Find the magnitude of a transfer function given by

$$H = \frac{4}{1 + j2\,\omega} \qquad\qquad \text{Equation 2-1}$$

1. The size of the numerator of the fraction is four.
2. The size of the denominator of the fraction is

$$\sqrt{(1)^2 + (2\,\omega)^2} \text{ or } \sqrt{1 + 4\,\omega^2}$$

3. The transfer function is then given by

$$H = \frac{4}{\sqrt{1 + 4\,\omega^2}}$$

As you can see, the function is related to frequency. At a frequency of $(\omega) = 0$, the function becomes

$$H = \frac{4}{\sqrt{1 + 4\,(0)^2}} = 4$$

As frequency increases the function decreases. At a frequency of infinity, the function becomes zero. Therefore, we can say we have a low-pass network. (A network which passes the lower frequencies with little attenuation, but passes the higher frequencies with large amounts of attenuation. The plot of a low-pass-filter magnitude response is shown in Fig. 2-1.

Another characteristic of a filter is the phase response. To determine the phase, proceed as given in the following example.

Example 2-2. What is the phase behavior of the transfer function given in Example 2-1?

Examination of the transfer functions shows that the numerator has a value of four. There is no imaginary part of the numerator. Using elementary trigonometry, we know that the phase of a number such as $C = a + jb$ is $\theta = \arctan \frac{b}{a}$. Since the imaginary part of the numerator has a size of zero, the phase of the numerator (θ_n) is given by the equation

$$\theta_n = \arctan \frac{0}{4} = 0°$$

The denominator also has a phase. Using the same procedure, we find

$$\theta_d = \arctan \frac{2\omega}{1} = \arctan 2\omega$$

The phase of the entire function is referenced to the numerator. The overall phase is the difference between the phase in the denominator and the phase in the numerator. The phase function is then

$$\theta = \theta_n - \theta_d$$
$$= 0 - \arctan 2\omega$$
$$= -\arctan 2\omega$$

The phase then has a value of zero at $\omega = 0$. As frequency increases the phase θ reaches $-90°$ when $\omega = \infty$. Note: $-\arctan(\infty) = -90°$.

Complete Transfer Functions

The complete transfer function has both phase and magnitude. If we call the magnitude $H(j\omega)$, and the phase θ, we could write the complete transfer function as

$$H(j\omega) = |H(j\omega)| \; \underline{/\theta}$$

Example 2-3. What is the complete transfer function of the problem from Example 2-1?

$$= |H(j\omega)| - \underline{/\theta} = \frac{4}{\sqrt{1 + 4\omega^2}} \; \underline{/- \arctan 2\omega}$$

Example 2-4. Determine the magnitude and phase for a transfer function of the form:

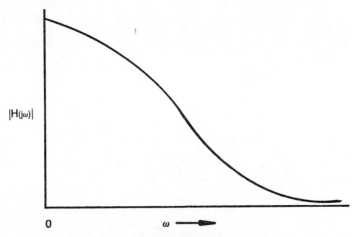

Fig. 2-1. Typical low-pass-filter magnitude response.

$$H(s) = \frac{1}{S^2 + \sqrt{2}\,S + 1}$$

First of all, you probably wonder what the S represents in the equation. It simply is another way of writing $j\,\omega$.

Replace the S and S^2 with $j\,\omega$ and $(j\,\omega)^2 = -\omega^2$ respectively:

$$H(j\omega) = \frac{1}{-W^2 + \sqrt{2}\,(j\omega) + 1}$$

Gather the real and imaginary parts in the denominator:

$$H(j\,\omega) = \frac{1}{1 - \omega^2 + j\,\sqrt{2}\,\omega}$$

Find the magnitude of $H(j\omega)$

$$|H(j\omega)| = \frac{1}{\sqrt{(1-\omega^2)^2 + [\sqrt{(2}\,\omega)]^2}} = \frac{1}{\sqrt{(1-2\omega^2 + \omega^4) + 2\omega^2}}$$

$$= \frac{1}{\sqrt{1 + \omega^4}}$$

Find the phase of $H(j\omega)$:

$\theta = $ phase of numerator - phase of denominator

$$\theta = \arctan\frac{0}{1} - \arctan\frac{\sqrt{2}\,\omega}{1 - \omega^2} \qquad \theta = -\arctan\frac{\sqrt{2}\,\omega}{1 - \omega^2}$$

The answer to the problem is then

$$H(j\omega) = \frac{\underline{/-\arctan} \quad \dfrac{\sqrt{2}\omega}{1 - \omega^2}}{\sqrt{1 + \omega^4}}$$

Example 2-5. Determine the magnitude and phase for a transfer function of the form

$$H(s) = \frac{6S + 3}{S^2 + 4S + 1}$$

$$H(j\omega) = \frac{6j\,\omega + 3}{-\omega^2 + 4j\omega + 1} = \frac{3 + 6j\omega}{1 - \omega^2 + 4j\omega}$$

$$H(j\omega) = \frac{\sqrt{(3)^2 + (16\,\omega^2)}}{\sqrt{(1 - \omega^2)^2 + (16\,\omega)^2}}$$

$$H(j\omega) = \frac{\sqrt{9 + 36\,\omega^2}}{\sqrt{1 - 2\omega^2 + \omega^4 + 16\omega^2}}$$

$$H(j\omega) = \frac{\sqrt{9 + 36\omega^2}}{\sqrt{1 + 14\omega^2 + \omega^4}}$$

θ = phase of numerator - phase of denominator

$$= \arctan\frac{6\omega}{3} - \arctan\frac{4\omega}{1-\omega^2}$$

$$= \arctan 2\omega - \arctan\frac{4\omega}{1-\omega^2}$$

Writing the transfer function, we then have

$$H(j\omega) = \frac{\sqrt{9 + 36\omega^2}}{1 + 14\omega^2 + \omega^4} \underline{/\arctan 2\omega - \arctan \quad \frac{4\omega}{1-\omega^2}}$$

TYPES OF TRANSFER FUNCTIONS

There are many types of transfer functions with which you must become familiar.

Low-Pass-Filter Transfer Function

A low-pass filter has a response that is high at low frequencies and is low at high frequencies. The general form of the transfer function is given by

$$H(s) = \frac{K}{S + \alpha} \qquad \text{Equation 2-2}$$

This has a magnitude of that given by

$$H(j\omega) = \frac{K}{\sqrt{\omega^2 + \alpha^2}} \qquad \text{Equation 2-3}$$

where α is called the "cutoff" frequency of the network. As frequency goes up, a place is reached where ω equals α. At that frequency the magnitude of the transfer function is

$$\frac{K}{\sqrt{2\alpha^2}} = \frac{K}{\sqrt{2}\,\alpha}$$

This value is $\frac{1}{\sqrt{2}}$ multiplied by the value of the function at zero frequency, which is $\frac{K}{\alpha}$. Figure 2-2 shows the magnitude response of such a filter. Another example of a low-pass transfer function is that given by

$$H(s) = \frac{K}{S^2 + \alpha S + b} \qquad \text{Equation 2-4}$$

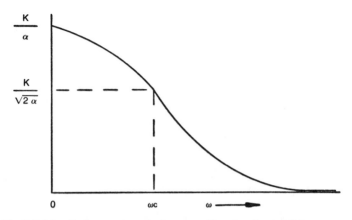

Fig. 2-2. Magnitude response of a low-pass filter showing cutoff frequency.

This function has a magnitude that falls more rapidly than the first example. As long as there is no S in the numerator, and some power S in the denominator, we know for sure that we have a low pass filter.

High-Pass-Filter-Transfer Function

A high-pass filter has a response that is low at low frequencies and is high at high frequencies. The general form of the transfer function is given by

$$H(s) = \frac{k S}{S + \alpha} \qquad \text{Equation 2-5}$$

where α is the cutoff frequency of the network. At zero frequency, the response is zero. At very high frequencies the transfer function approaches K. This can be seen by dividing numerator and denominator by S and then taking the limit as S approaches ∞.

$$H(s) = \frac{K}{\dfrac{S}{S} + \dfrac{\alpha}{S}} = \frac{K}{1 + \dfrac{\alpha}{S}}$$

$$H(s) = \frac{K}{1 + \dfrac{\alpha}{\infty}} = K$$

$$S \to \infty$$

K is a simple constant of the function. It can be any value between $-\infty$ and ∞. There is a frequency when the transfer function reaches $\dfrac{K}{\sqrt{2}}$. This is the cutoff frequency. It is simply "α" in this example. Figure 2-3 illustrates a typical high-pass characteristic.

Another example of a high-pass transfer function is that given by

$$H(s) = \frac{K S^2}{S^2 + \alpha S + b}$$ Equation 2-6

where K, α, and b are constants of the system.

Band-Pass Filter

A bandpass filter has a magnitude response as shown in Fig. 2-4. The response is low at low frequencies. It then rises as frequency increases and reaches a maximum at some frequency. After this, the response falls again and reaches zero at infinite frequencies. The transfer function takes the form

$$H(s) = \frac{K S}{S^2 + \alpha S + b}$$ Equation 2-7

It can be easily seen that if we change S to $j\omega$ we have

$$H(s) = \frac{K (j\omega)}{-\omega^2 + j\omega\alpha + b}$$ Equation 2-8

If ω equals zero, then

$$H(j\omega) = \frac{K j(0)}{-(0)^2 + j0\,(\alpha) + b} = 0$$

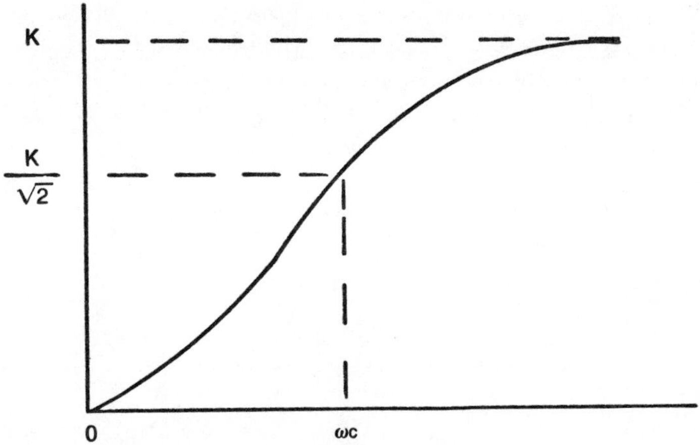

Fig. 2-3. Typical high-pass filter magnitude response.

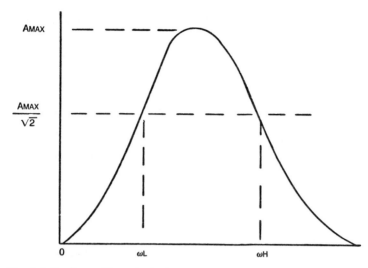

Fig. 2-4. Band-pass filter magnitude response.

If ω equals infinity, then

$$H(j\omega) = \left.\frac{K\dfrac{j\omega}{j\omega}}{-\dfrac{\omega^2}{j\omega}+\dfrac{j\omega\alpha}{j\omega}+\dfrac{b}{j\omega}}\right|_{\omega=\infty} = \left.\frac{K}{j\omega+\alpha-j\dfrac{b}{\omega}}\right|_{\omega}$$

$$= \infty = \frac{K}{j\infty+\alpha-j\dfrac{b}{\infty}} = \frac{K}{j\infty+\alpha-j(0)} = 0$$

There are two cutoff frequencies for each function. There is a low cutoff frequency (ω_L) and a high cutoff frequency (ω_H). The low cutoff frequency is that frequency, as you advance towards the peak of the magnitude response, where the magnitude will be 70.7% of the peak of the magnitude response. The high cutoff frequency is that frequency where the magnitude has fallen to 70.7% of the peak of the magnitude response. The difference between ω_H and ω_L is called the bandwidth.

Band-Reject Filter

A band-reject filter has a response in magnitude as shown in Fig. 2-5. The transfer function is as given in

$$H(s) = \frac{S^2 + C}{S^2 + \alpha S + b}$$

Equation 2-9

α, b and c are constants of the system. The magnitude of the response is given by

$$|H(j\omega)| = \frac{\sqrt{(-\omega^2 + C)^2}}{\sqrt{(-\omega^2 + b)^2 + (\alpha\omega)^2}}$$ Equation 2-10

At zero frequency, the magnitude has a response of

$$H(j\omega) = \frac{c}{b}$$

As frequency rises, the numerator begins to approach zero. At a value of frequency $\omega = \sqrt{c}$, the numerator goes to zero. The magnitude is then zero. As frequency continues to advance, the numerator gets larger again and the magnitude of the transfer function goes up. Finally, at $\omega = \infty$, the magnitude goes to unity. This can be seen because:

$$H(j\omega) = \frac{-\omega^2 + c}{-\omega^2 + j\omega\alpha + b} \Bigg|_{\omega = \infty} = \frac{\dfrac{-\omega^2}{\omega^2} + \dfrac{c}{\omega^2}}{\dfrac{-\omega^2}{\omega^2} + \dfrac{j\omega\alpha}{\omega^2} + \dfrac{b}{\omega^2}} \Bigg|_{\omega = \infty}$$

$$= \frac{-1 + \dfrac{C}{\omega^2}}{-1 + j\dfrac{\alpha}{\omega} + \dfrac{b}{\omega^2}} \Bigg|_{\omega = \infty}$$

$$= \frac{-1 + \dfrac{c}{(\infty)^2}}{-1 + \dfrac{j\alpha}{(\infty)} + \dfrac{b}{(\infty)^2}}$$

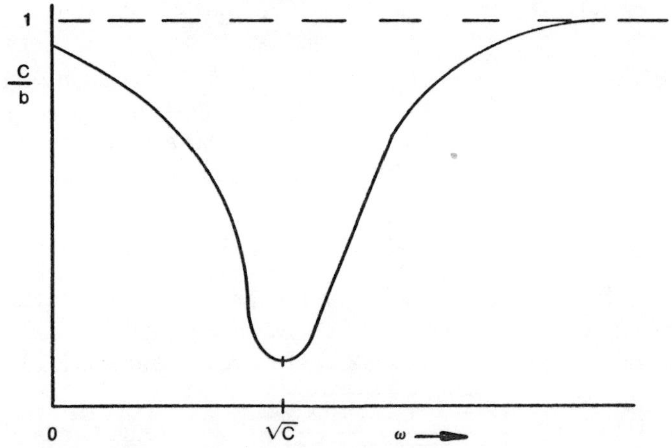

Fig. 2-5. Band-rejection filter magnitude response.

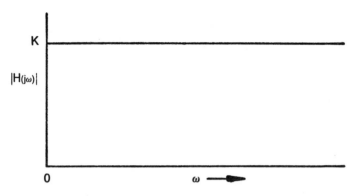

Fig. 2-6. All-pass filter magnitude response.

$$= \frac{-1 + 0}{-1 + j0 + 0} = 1$$

All-Pass Filter

An all pass filter has the response shown in Fig. 2-6. The basic equations may take various forms. Four of the equations are given in equations 2-11 through 2-14.

$$H(s) = \frac{s + \alpha}{s - \alpha} \qquad \text{Equation 2-11}$$

$$H(s) = \frac{s - \alpha}{s + \alpha} \qquad \text{Equation 2-12}$$

$$H(s) = \frac{S^2 + \alpha S + b}{S^2 - \alpha S + b} \qquad \text{Equation 2-13}$$

$$H(s) = \frac{S^2 - \alpha S + b}{S^2 + \alpha S + b} \qquad \text{Equation 2-14}$$

Let us look at the magnitudes of each of these functions. The magnitude of equation 2-11 is given by equation 2-15.

$$|H(j\omega)| = \frac{\sqrt{\omega^2 + \alpha^2}}{\sqrt{\omega^2 + \alpha^2}} \qquad \text{Equation 2-15}$$

For any frequency ω, the value of $|H(j\omega)|$ is always 1. The magnitude of equation 2-12 is the same as equation 2-15. The magnitude of equations 2-13 and 2-14 are given by equation 2-16.

$$|H(j\omega)| = \frac{\sqrt{(-\omega^2 + b)^2 + (\alpha\omega)^2}}{\sqrt{(-\omega^2 + b)^2 + (\omega)^2}} \qquad \text{Equation 2-16}$$

R

Fig. 2-7. A simple resistor of value R.

Again, you can see that for every value of ω the value of $|H(j\omega)|$ is unity. It may seem rather strange to have a transfer function that has a magnitude of unity for all frequencies. Why go to the trouble of building a network that produces the same output as the input? The answer is that the difference comes in regards to the phase. The phase is not constant, but varies with frequency. We will learn more about the all-pass filter network later in this book.

IMPEDANCE NETWORKS

The student of network synthesis should become familiar with various impedance equations and their respective circuits. For example, if we write $Z = R$ it represents a simple resistor of value R, see Fig. 2-7. Anyone with some understanding of basic electricity realizes this. Now let us look at some other examples.

Example 2-6. Given the impedance equation

$$Z = R + j\omega L$$
$$= R + SL$$

Equation 2-17

This represents the circuit of Fig. 2-8. A simple resistance R in series with an inductor of inductance value L.

Example 2-7. Given the impedance equation 2-18. What circuit does it represent?

$$Z = \frac{j\omega CR + 1}{j\omega C}$$

Equation 2-18

$$= \frac{SCR + 1}{SC} = R + \frac{1}{SC}$$

This represents the circuit of Fig. 2-9.

Example 2-8. Now let's get a little more difficult. Given the equation 2-19, what is the circuit it represents?

$$Z = \frac{j\omega L}{1 - \omega^2 LC}$$

Equation 2-19

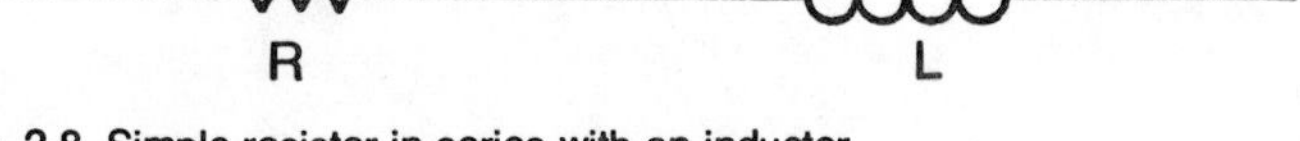

R   L

Fig. 2-8. Simple resistor in series with an inductor.

Fig. 2-9. Simple resistor in series with a capacitor.

First of all, let us see if we can represent the equation by a product of two impedances.

Divide numerator and denominator of the equation by C

$$Z = \frac{j\omega\dfrac{L}{C}}{\dfrac{1}{C} - \dfrac{\omega^2 LC}{C}} \qquad \text{Equation 2-20}$$

Divide numerator and denominator of the equation by $j\omega$.

$$Z = \frac{\dfrac{j\omega L}{j\omega C}}{\dfrac{1}{j\omega C} - \dfrac{\omega^2 LC}{C(j\omega)}} \qquad \text{Equation 2-21}$$

Equation 2-23 can be rewritten as the product over the sum of two impedances.

$$Z = \frac{j\omega L\left(\dfrac{1}{j\omega C}\right)}{\dfrac{1}{j\omega C} \quad \dfrac{\omega L}{j}} = \frac{(j\omega L)\left(\dfrac{1}{j\omega C}\right)}{\dfrac{1}{j\omega C} - (-j)(\omega L)}$$

$$Z = \frac{(j\omega L)\left(\dfrac{1}{j\omega C}\right)}{\dfrac{1}{j\omega C} + j\omega L} \qquad \text{Equation 2-22}$$

Equation 2-24 represents a capacitive reactance in parallel with an inductive reactance. See Fig. 2-10 for the effective circuit.

Example 2-9. Given the impedance relation given as

$$Z = \frac{-\omega^2 LC + 1}{j\omega C} \qquad \text{Equation 2-23}$$

What is the circuit this equation represents?

$$Z = \frac{-\omega^2 LC}{j\omega C} + \frac{1}{j\omega C} \qquad \text{Equation 2-24}$$

$$= j\omega L + \frac{1}{j\omega c}$$

48

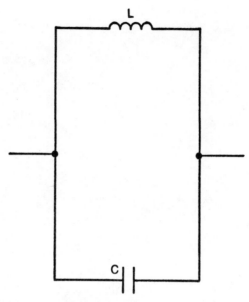

Fig. 2-10. Capacitor and inductor in parallel.

This is simply the impedance of an inductor of value L and a capacitor of value C in series. The effective circuit is shown in Fig. 2-11.

Example 2-10. What circuit does equation 2-25 represent?

$$Z = \frac{j\omega CR - \omega^2 LC + 1}{j\omega C}$$

$$Z = R + \frac{-\omega L}{j} + \frac{1}{j\omega C} \qquad \text{Equation 2-25}$$

$$Z = R + j\omega L + \frac{1}{j\omega C}$$

Figure 2-12 shows the effective circuit composed of a resistor, inductor, and capacitor in series.

Example 2-11. Equation 2-26 represents the impedance of a certain circuit. Draw the circuit.

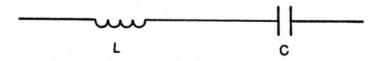

Fig. 2-11. Capacitor and inductor in series.

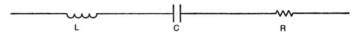

Fig. 2-12. Series circuit composed of a resistor, inductor, and capacitor.

$$Z = \frac{R}{1 + j\omega CR} \qquad \text{Equation 2-26}$$

$$Z = \frac{\dfrac{R}{j\omega C}}{\dfrac{1}{j\omega C} + R}$$

$$Z = \frac{R\left(\dfrac{1}{j\omega C}\right)}{R + \dfrac{1}{j\omega C}}$$

This represents a resistor of value R and a capacitor of value C in parallel. Figure 2-13 illustrates this equation.

Example 2-12. What circuit is represented by equation 2-27?

$$Z = \frac{j\omega LR}{j\omega L + R} \qquad \text{Equation 2-27}$$

By simple inspection, we see this is simply an inductance L and resistance R in parallel as shown in Fig. 2-14.

Example 2-13. What circuit is represented by equation 2-28?

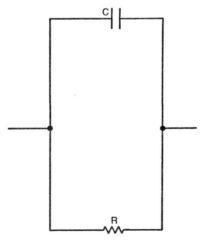

Fig. 2-13. Circuit representing Equation 2-26.

$$Z = \frac{R + j\omega L}{j\omega CR - \omega^2 LC + 1} \qquad \text{Equation 2-28}$$

If we divide numerator and denominator by C, we have

$$Z = \frac{\dfrac{R + j\omega L}{j\omega C}}{\dfrac{j\omega CR}{j\omega C} - \dfrac{\omega^2 LC}{j\omega C} + \dfrac{1}{j\omega C}} \qquad \text{Equation 2-29}$$

$$Z = \frac{(R + j\omega L)\left(\dfrac{1}{j\omega C}\right)}{R + j\omega L + \dfrac{1}{j\omega C}} \qquad \text{Equation 2-30}$$

Equation 2-30 represents two branches in parallel. An impedance of $R + j\omega L$ is in one branch and an impedance of $\dfrac{1}{j\omega C}$ in the other branch. Figure 2-15 represents the circuit.

CHAPTER SUMMARY

In this chapter, we have covered some of the basic problems of network synthesis. I cannot emphasize enough that you become

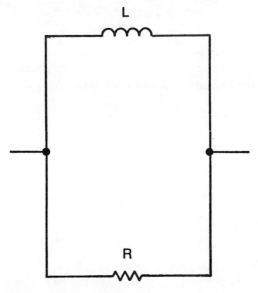

Fig. 2-14. Circuit representing Equation 2-27.

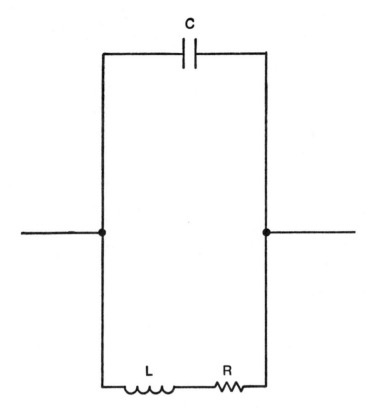

Fig. 2-15. Inductor and resistor in series, in parallel with a capacitor.

familiar with transfer-function equations. You must be able to examine an equation and determine what type of circuit it represents.

Problems

1. Draw the circuits that are represented by the following equations:

a. $Z(s) = \dfrac{3}{1 + 16S} + \dfrac{1 + 48S + 64S^2}{16S}$

b. $Z(s) = \dfrac{19S}{1 + 42S^2} + \dfrac{1 + 32S^2}{4S}$

2. Which of the following impedances cannot be produced with a combination of components such as inductors, capacitors, and resistors? Why?

a. $Z(j\omega) = -4 + j6$

b. $Z(j\omega) = 6j\omega + \dfrac{1}{3\,j\omega}$

c. $Z(j\omega) = 6j\omega + j7\omega^2$

3. Determine the magnitude of the following functions and make a stretch of the magnitude versus frequency for $\omega = 0,1,2,3,4,$ and 5 radians.

a. $H(s) = \dfrac{4}{S^2 + 3S + 2}$

b. $H(s) = \dfrac{S^2 + 4}{S^2 + 3S + 2}$

c. $H(s) = \dfrac{4S}{S^2 + 3S + 2}$

d. $H(s) = \dfrac{S^2 + 3S + 2}{S^2 - 3S + 2}$

e. $H(s) = \dfrac{4S^2}{S^2 + 3S + 2}$

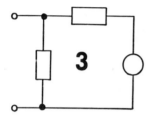

Deriving Transfer Functions

Now let us investigate some more advanced circuits and derive their transfer functions by working many examples.

Example 3-1. What is the transfer function for the circuit of Fig. 3-1?

Using the voltage-divider principle,

$$\frac{E_o}{E_i} = \frac{\dfrac{1}{SC}}{R + \dfrac{1}{SC}}$$

$$\frac{E_o}{E_i} = \frac{1}{SCR + 1}$$

$$\frac{E_o}{E_i} = \frac{1}{j\omega CR + 1}$$

Example 3-2. What is the transfer function for the circuit of Fig. 3-2?

Using mesh equations, we write

$$E_i = I_1\left(R_1 + \frac{1}{C_1 S}\right) - I_2\left(\frac{1}{C_1 S}\right)$$

$$0 = -I_1\left(\frac{1}{C_1 S}\right) + I_2\left(\frac{1}{C_1 S} + R_2 + \frac{1}{C_2 S}\right)$$

Since $R_1 = R_2 = R$, and $C_1 = C_2 = C$, we may write

$$E_i = I_1\left(R + \frac{1}{CS}\right) - I_2\left(\frac{1}{CS}\right)$$

$$0 = -I_1\left(\frac{1}{CS}\right) + I_2\left(\frac{1}{CS} + R + \frac{1}{CS}\right)$$

Solving for I_2, we first determine Δ of the system.

$$\begin{aligned}
\Delta &= \left| R + \frac{1}{CS}, -\frac{1}{CS} - \frac{1}{CS}, \frac{2}{CS} + R \right. \\
&= \left(R + \frac{1}{CS}\right)\left(\frac{2}{CS} + R\right) - \left(\frac{-1}{CS}\right)\left(\frac{-1}{CS}\right) \\
&= \frac{2R}{CS} + R^2 + \frac{2}{(CS)^2} + \frac{R}{CS} - \frac{1}{(CS)^2} = \frac{3R}{CS} + R^2 + \frac{1}{(CS)^2}
\end{aligned}$$

Solving for I_2, we write

$$\begin{aligned}
\Delta I_2 &= \left| R + \frac{1}{CS}, \ E_1 - \frac{1}{CS}, \ 0 \right. \\
&= \left(R + \frac{1}{CS}\right)(0) - \left(\frac{-1}{CS}\right)E_1 = \frac{E_i}{CS}
\end{aligned}$$

To determine the output voltage, we can write

$$E_o = I_2\left(\frac{1}{SC_2}\right)$$

Substituting for I_2, we obtain

$$E_o = \frac{\Delta I_2}{\Delta}\left(\frac{1}{SC}\right) = \frac{\dfrac{(E_i)}{CS}\left(\dfrac{1}{SC}\right)}{\dfrac{3R}{CS} + R^2 + \dfrac{1}{(CS)^2}}$$

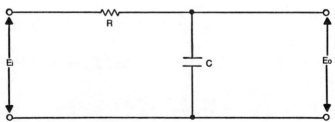

Fig. 3-1. Circuit for example 3-1.

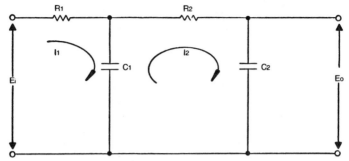

Fig. 3-2. Circuit for example 3-2.

$$E_o = \frac{\frac{E_i}{(CS)^2}}{\frac{3R}{CS} + R^2 + \frac{1}{(CS)^2}}$$

Multiplying by $(CS)^2$, we obtain

$$E_o = \frac{E_i}{3\,RCS + (CSR)^2 + 1}$$

Divide both sides of the equation by E_i and rearrange

$$\frac{E_o}{E_i} = \frac{1}{1 + 3SCR + (SCR)^2}$$

Example 3-3. What is the transfer function for the circuit of Fig. 3-3?

Using mesh equations, we have

$$E_i = I_1 \left(R_1 + \frac{1}{C_1 S}\right) - I_2 \frac{(1)}{C_1 S}$$

$$0 = -I_1 \frac{(1)}{C_1 S} + I_2 \left(\frac{1}{C_1 S} + R_2 + \frac{1}{C_2 S}\right) - I_3 \left(\frac{1}{CS}\right)$$

$$0 = -I_2 \left(\frac{1}{C_2 S}\right) + I_3 \left(\frac{1}{C_2 S} + R_3 + \frac{1}{C_3 S}\right)$$

Since $R_1 = R_2 = R_3 = R$, and $C_1 = C_2 = C_3 = C$, we can write the following equations:

$$E_i = I_1 \left(R + \frac{1}{CS}\right) - I_2 \left(\frac{1}{CS}\right)$$

$$0 = -I_1\left(\frac{1}{CS}\right) + I_2\left(\frac{1}{CS} + R + \frac{1}{CS}\right) - I_3\left(\frac{1}{CS}\right)$$

$$0 = -I_2\left(\frac{1}{CS}\right) + I_3\left[\left(\frac{1}{CS}\right) + R + \left(\frac{1}{CS}\right)\right]$$

Since the output voltage is dependent on the product of I_3 and the reactance of C_3, we can write,

$$E_o = I_3\frac{1}{CS}$$

To solve for I_3, we first determine the Δ of the system:

$$\Delta = \begin{vmatrix} R+\frac{1}{CS}, -\frac{1}{CS}, 0 - \frac{1}{CS}, \frac{2}{CS}+ R, -\frac{1}{CS} \\ \\ 0, -\frac{1}{CS}, \frac{2}{CS}+ R \end{vmatrix}$$

$$\Delta = \left(R+\frac{1}{CS}\right)\left[\left(\frac{2}{CS}+ R\right)^2 - \left(-\frac{1}{CS}\right)^2\right]$$

$$- \left(-\frac{1}{CS}\right)\left[\left(-\frac{1}{CS}\right)\left(\frac{2}{CS}+ R\right)\right]$$

$$= \left(R+\frac{1}{CS}\right)\left[\frac{3}{C^2S^2}+ \frac{4R}{CS} + R^2\right]-\frac{2}{C^3 S^3} - \frac{R}{C^2 S^2}$$

$$= \frac{3R}{C^2 S^2} + \frac{4R^2}{CS} + R^3 + \frac{3}{C^3 S^3} + \frac{4R}{C^2 S^2}+ \frac{R^2}{CS} -\frac{2}{C^3 S^3}$$

$$- \frac{R}{C^2S^2} = \frac{6R}{C^2S^2} + \frac{5R^2}{CS} + R^3 + \frac{1}{C^3S^3}$$

Solving for ΔI_3 we have:

$$\Delta I_3 = \begin{vmatrix} R + \frac{1}{CS}, -\frac{1}{CS}, E_i - \frac{1}{CS}, \frac{2}{CS} + R, 0 \\ \\ 0, -\frac{1}{CS}, 0 = E_i \left(\frac{1}{CS}\right)^2 = \frac{E_i}{(CS)^2} \end{vmatrix}$$

Writing the equation for E_o, we obtain

$$E_o = \frac{\Delta I_3}{\Delta} \left(\frac{1}{C_3S}\right) = \frac{\dfrac{E_i}{(CS)^2}\left(\dfrac{1}{(CS)}\right)}{\dfrac{6R}{C^2 S^2} + \dfrac{5R^2}{CS} + R^3 + \dfrac{1}{C^3S^3}}$$

57

By multiplying numerator and denominator by $(CS)^3$, we obtain:

$$E_o = \frac{E_i}{6R\,C\,S + 5R^2\,C^2\,S^2 + R^3\,C^3\,S^3 + 1}$$

Dividing both sides of the equation by E_i and arranging:

$$\frac{E_o}{E_i} = \frac{1}{1 + 6\,S\,C\,R + 5\,(SCR)^2 + (SCR)^3}$$

Example 3-4. Write the phase equations and magnitude equations from the transfer functions derived in examples 3-2 and 3-3.

For the transfer function

$$\frac{E_o}{E_i} = \frac{1}{1 + 3\,S\,C\,R + (SCR)^2}$$

Let $S = j\omega$

$$\frac{E_o}{E_i} = \frac{1}{1 + 3j\omega CR + (j\omega CR)^2}$$

$$= \frac{1}{1 + 3j\omega CR - \omega^2\,C^2\,R^2}$$

The magnitude becomes

$$\frac{E_o}{E_i} = \frac{1}{\sqrt{(1 - \omega^2 C^2 R^2)^2 + (3\omega CR)^2}}$$

$$= \frac{1}{\sqrt{1 - 2\omega^2\,C^2\,R^2 + \omega^4\,C^4\,R^4 + 9\omega^2\,C^2\,R^2}}$$

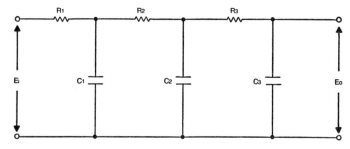

Fig. 3-3. Circuit for example 3-3.

$$= \frac{1}{\sqrt{1 + 7\,\omega^2\,C^2\,R^2 + \omega^4\,C^4\,R^4}}$$

The phase equation becomes,

$$\theta = \theta_n - \theta_d = \arctan\frac{0}{1} - \arctan\frac{3\,\omega CR}{1 - \omega^2 C^2 R^2}$$

$$= -\arctan\frac{3\,\omega\,CR}{1 - \omega^2 C^2 R^2}$$

For the transfer function,

$$\frac{E_o}{E_i} = \frac{1}{1 + 6SCR + 5\,(SCR)^2 + (SCR)^3}$$

Let $j\omega = S$

$$\frac{E_o}{E_i} = \frac{1}{1 + 6j\,\omega CR + 5\,(j\,\omega CR)^2 + (j\omega CR)^3}$$

$$= \frac{1}{1 + 6j\,\omega CR - 5\,\omega^2\,C^2\,R^2 - j\,\omega^3\,C^3\,R^3}$$

$$\frac{E_o}{E_i} = \frac{1}{\sqrt{(1 - 5\,\omega^2\,C^2\,R^2)^2 + (6\,\omega CR - \omega^3\,C^3\,R^3)^2}} =$$

$$\frac{1}{\sqrt{1 - 10\,\omega^2 C^2 R^2 + 25\,\omega^4\,C^4\,R^4 + 36\,\omega^2\,C^2\,R^2 - 12\,\omega^4\,C^4\,R^4 + \omega^6 C^6\,R^6}}$$

$$\frac{E_o}{E_i} = \frac{1}{\sqrt{1 + 26\,\omega^2\,C^2\,R^2 + 13\,\omega^4\,C^4\,R^4 + \omega^6\,C^6\,R^6}}$$

The phase equation is

$$\theta = \theta_n - \theta_d = \arctan\frac{\theta}{1} - \arctan\frac{6\,\omega CR - \omega^3\,C^3\,R^3}{1 - 5\,\omega^2\,C^2\,R^2}$$

$$= -\arctan\frac{6\,\omega CR - \omega^3\,C^3\,R^3}{1 - 5\,\omega^2\,C^2\,R^2}$$

Example 3-5. Find the magnitude and phase equations for the following transfer function:

$$H = \frac{1}{1 + j\,\omega CR}$$

The magnitude becomes

$$H(j\omega) = \frac{1}{\sqrt{1 + (\omega CR)^2}}$$

The phase equation becomes

$$\theta = \theta n - \theta d$$
$$= \arctan\frac{\theta}{1} - \arctan\frac{\omega CR}{1} = -\arctan\omega CR$$

We now have evaluated three transfer functions; one for a single-section RC low-pass filter, one for a double-section RC low-pass filter, and one for a triple-section RC low-pass filter. Suppose we plug in some values and check each response at several frequencies.

For simplicity, let R = 1 ohm and C = 1 farad.

The three transfer functions become

One section:

$$H(j\omega) = \frac{1}{1 + j\omega}$$

Two sections:

$$H(j\omega) = \frac{1}{1 + 3j\omega - \omega^2}$$

Three sections:

$$H(j\omega) = \frac{1}{1 + 6j\omega - 5\omega^2 - j\omega^3}$$

The magnitude for each section, and the phase equations are

One section:

$$|H(j\omega)| = \frac{1}{\sqrt{1 + \omega^2}}$$
$$\theta = -\arctan\omega$$

Two sections:

$$|H(j\omega)| = \frac{1}{\sqrt{1 + 7\omega^2 + \omega^4}}$$
$$\theta = -\arctan\frac{3\omega}{1 - \omega^2}$$

60

Three sections:

$$|H(j\omega)| = \frac{1}{\sqrt{1 + 26\omega^2 + 13\omega^4\,\omega + ^6}}$$

$$\theta = -\arctan\frac{6\omega - \omega^3}{1 - 5\omega^2}$$

If we evaluate the phase and magnitude for each network at ω = 0, 1, 2, 3, 10, 100, and ∞ radians we have:

One section:

$\omega = 0$

$$|H(j\omega)| = \frac{1}{\sqrt{1 + \omega^2}} = \frac{1}{\sqrt{1}} = 1$$

$$\theta(\omega) = -\arctan\omega = -\arctan 0 = 0°$$

$\omega = 1$

$$|H(j\omega)| = \frac{1}{\sqrt{1 + (1)^2}} = \frac{1}{\sqrt{2}} = 0.707$$

$$\theta(\omega) = -\arctan 1 = -45°$$

$\omega = 2$

$$|H(j\omega)| = \frac{1}{\sqrt{1 + (2)^2}} = \frac{1}{\sqrt{5}} = 0.447$$

$$\theta(\omega) = \arctan - 2 = 63.4°$$

$\omega = 3$

$$|H(j\omega)| = \frac{1}{\sqrt{1 + (3)^2}} = \frac{1}{\sqrt{10}} = .316$$

$$\theta(\omega) = -\arctan 3 = -71.5°$$

$\omega = 10$

$$|H(j\omega)| = \frac{1}{\sqrt{1 + (10)^2}} = \frac{1}{\sqrt{101}} = .099$$

$$\theta(\omega) = -\arctan 10 = -84.3°$$

$\omega = 100$

$$|H(j\omega)| = \frac{1}{\sqrt{1 + (100)^2}} = .01$$

61

$$\theta(\omega) = -\arctan 100 = -89.4°$$

$$\omega = \infty$$

$$H(j\omega) = \frac{1}{\sqrt{1 + (\infty)^2}} = 0$$

$$\theta(\omega) = -\arctan \infty = -90°$$

Two sections:

$$\omega = 0$$

$$|H(j\omega)| = \frac{1}{\sqrt{1 + 7\omega^2 + \omega^4}}$$

$$|H(j\omega)| = \frac{1}{\sqrt{1 + 0 + 0}} = 1$$

$$\theta = -\arctan \frac{3\omega}{1 - \omega^2} = -\arctan \frac{3(0)}{1 - 0^2} = 0°$$

$$\omega = 1$$

$$|H(j\omega)| = \frac{1}{\sqrt{1 + 7 + 1}} = \frac{1}{\sqrt{9}} = \frac{1}{3} = 0.333$$

$$\theta = -\arctan \frac{3}{1 - 1^2} = -\arctan \frac{3}{0}$$

$$= -\arctan \infty = -90°$$

$$\omega = 2$$

$$|H(j\omega)| = \frac{1}{\sqrt{1 + 7(2)^2 + (2)^4}} = \frac{1}{\sqrt{1 + 28 + 16}}$$

$$H(j\omega) = \frac{1}{45} = 0.149$$

$$\theta = -\arctan \frac{3(2)}{1 - 2^2} = -\arctan \frac{6}{-3}$$

$$\theta = -110.6$$

$$\omega = 3$$

$$|H(j\omega)| = \frac{1}{\sqrt{1 + 7(3)^2 + (3)^4}} = \frac{1}{\sqrt{1 + 63 + 81}}$$

$$= \frac{1}{\sqrt{145}} = .083 \quad \theta = -\arctan \frac{3\ (3)}{1 - 3^2}$$

$$= -\arctan \frac{9}{-8} = -131.7°$$

$\omega = 10$

$$|H(j\ \omega)| = \frac{1}{\sqrt{1 + 7\ (10)^2 + (10)^4}}$$

$$= \frac{1}{\sqrt{1 + 700 + 10,000}}$$

$$|H(j\ \omega)| = .0096$$

$$\theta = -\arctan \frac{3\ (10)}{1 - (10)^2}$$

$$= -\arctan \frac{30}{-99} = -163.2°$$

$\omega = \infty$

$$|H(j\ \omega)| = \frac{1}{\sqrt{1 + 7\ (\infty)^2 + (\infty)^4}} = 0$$

$$\theta = -\arctan \frac{3\ (\infty)}{1 - (\infty)^2} = -180°$$

Three sections:

$$|H(j\ \omega)| = \frac{1}{\sqrt{1 + 26\ \omega^2 + 13\ \omega^4 + \omega^6}}$$

$\omega = 0$

$$H(j\ \omega) = \frac{1}{\sqrt{1 + 26\ (0)^2 + 13\ (0)^4 + (0)^6}}$$

$$= \frac{1}{\sqrt{1}} = 1$$

$$\theta = -\arctan \frac{6\omega - \omega^3}{1 - 5\omega^2}$$

$$= -\arctan \frac{6\ (0) - (0)^3}{1 - 5\ (0)^2} = 0°$$

$\omega = 1$

$$|H(j\omega)| = \frac{1}{\sqrt{1 + 26\,(1)^2 + 13\,(1)^4 + (1)^6}}$$

$$= \frac{1}{\sqrt{1 + 26 + 13 + 1}}$$

$$= \frac{1}{\sqrt{41}} = 0.156$$

$$\theta = -\arctan\frac{6\,(1) - (1)^3}{1 - 5\,(1)^2} = -\arctan\frac{6-1}{1-5}$$

$$= -\arctan\frac{5}{-4} = -128.7°$$

$\omega = 2$

$$|H(j\omega)| = \frac{1}{\sqrt{1 + 26\,(2)^2 + 13\,(2)^4 + (2)^6}}$$

$$= \frac{1}{\sqrt{1 + 104 + 208 + 64}}$$

$$= \frac{1}{\sqrt{377}} = .0515$$

$$\theta = -\arctan\frac{12 - 8}{1 - 20}$$

$$= -\arctan\frac{4}{-19} = -168.2°$$

$\omega = 3$

$$H(j\omega) = \frac{1}{\sqrt{1 + 26\,(3)^2 + 13\,(3)^4 + (3)^6}}$$

$$= \frac{1}{\sqrt{1 + 234 + 1053 + 279}}$$

$$= \frac{1}{\sqrt{2017}} = .022$$

$$\theta(\omega) = -\arctan\frac{6\,(3) - (3)^3}{1 - 5\,(3)^2}$$

$$= -\arctan\frac{18 - 27}{1 - 45}$$

$$= -\arctan\frac{-9}{-44} = -191.6°$$

64

$$\omega = 10$$

$$|H(j\omega)| = \frac{1}{\sqrt{1 + 26\,(10)^2 + 13\,(10)^4 + (10)^6}}$$

$$= \frac{1}{\sqrt{1 + 2600 + 130000 + 10^6}}$$

$$H(j\omega) = \frac{1}{\sqrt{1,132,601}} = 0.00093$$

$$\theta(\omega) = -\arctan\frac{6\,(10) - (10)^3}{1 - 5\,(10)^2}$$

$$= -\arctan\frac{60 - 1000}{1 - 500}$$

$$= -\arctan\frac{-940}{-499} = -242°$$

$$\omega = \infty$$

$$|H(j\omega)| = \frac{1}{\sqrt{1 + 26\,(\infty)^2 + 13\,(\infty)^4 + (\infty)^6}} = 0$$

$$\theta(\omega) = -\arctan\frac{6\,(\infty) - (\infty)^3}{1 - 5\,(\infty)^2}$$

$$= -\arctan\frac{-(\infty)^2}{-\infty} = -270°$$

We have now evaluated three different networks composed of identical values of C and R. If we plot the magnitude and phase versus frequency, we have the frequency response as shown in Fig. 3-4. Note that at 1 radian the one-section RC network has a magnitude of 0.707, while at the same frequency the two-section RC network has a magnitude of 0.333. Likewise, the three-section network response at 1 radian per second is 0.156. Note that the phase between output and input is respectively -45, -90 and $-128.7°$ at 1 radian per second.

Example 3-5. Derive the transfer function for the circuit in Fig. 3-5.

Again, we write the mesh equations:

$$E_i = I_1\,(SL + R_1) - I_2\,(R_1)$$

$$0 = -I_1\,R_1 + I_2\,(R_1 + \frac{1}{CS} + R_2)$$

We then determine the "delta" of the system:

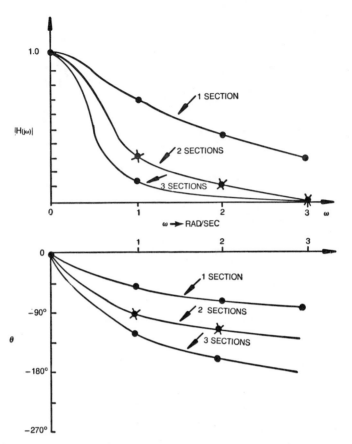

Fig. 3-4. Magnitude and phase for examples 3-1, 3-2, and 3-3.

$$\Delta = \begin{vmatrix} SL + R_1, & -R_1 \\ -R_1, & R_1 + R_2 + \dfrac{1}{CS} \end{vmatrix}$$

$$= (SL + R_1)(R_1 + R_2 + \frac{1}{CS}) - (-R_1)(-R_1)$$

$$= SLR_1 + SLR_2 + \frac{L}{C} + R_1^2 + R_1 R_2 + \frac{R_1}{CS} - R_1^2$$

$$= SLR_1 + SLR_2 + \frac{L}{C} + R_1 R_2 + \frac{R_1}{CS}$$

Then find the quantity ΔI_2:

$$\Delta I_2 = \begin{vmatrix} SL + R_1, & E_i & -R_1, & 0 \end{vmatrix}$$

66

$$\Delta I_2 = -(-R_1)(E_i) = E_i R_1$$

Writing the equation for the output voltage, we have

$$
\begin{aligned}
E_o &= I_2 R_2 \\
&= \frac{\Delta I_2}{\Delta} R_2 \\
&= \frac{E_i R_1}{SLR_1 + SLR_2 + \dfrac{L}{C} + R_1 R_2 + \dfrac{R_1}{CS}} R_2 \\
&= \frac{E_i R_1 R_2}{SLR_1 + SLR_2 + \dfrac{L}{C} + R_1 R_2 + \dfrac{R_1}{CS}}
\end{aligned}
$$

Multiplying by CS, we have

$$E_o = \frac{E_i R_1 R_2 CS}{S^2 LCR_1 + S^2 LCR_2 + LS + R_1 R_2 CS + R_1}$$

Dividing both sides by E_i, we obtain

$$\frac{E_o}{E_i} = \frac{R_1 R_2 CS}{S^2 LC(R_1 + R_2) + S(L + R_1 R_2 C) + R_1}$$

Placing the component values in the transfer function,

$$\frac{E_o}{E_i} = \frac{(2)(1)(1)S}{S^2(1)(1)(1+2) + S\left(1 + (2)(1)(1)\right) + 2}$$

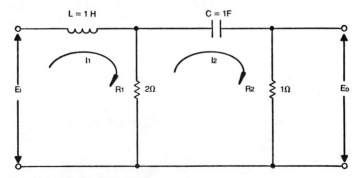

Fig. 3-5. Circuit for example 3-5.

$$\frac{E_o}{E_i} = \frac{2 \; S}{S^2 + S(1+2) + 2} = \frac{(2)S}{S^2 + 3S + 2}$$

Multiplying the numerator and denominator by 3, we obtain

$$\frac{E_o}{E_i} = \frac{2\,S}{3S^2 + 9S + 6}$$

It is interesting to note that the denominator of this fraction could be factored into $(3S + 3)(S + 2)$. We could then write the following equation:

$$\frac{E_o}{E_i} = \frac{2S}{(3S + 3)(S + 2)}$$

$$\frac{E_o}{E_i} = \frac{2\,j\omega}{(3\,j\omega + 3)(j\omega + 2)}$$

TIME DELAY

From the study of physics, we know that angular distance is the product of frequency and time. In other words we have,

$$\theta = \omega t \qquad\qquad \text{Equation 3-1}$$

where θ is the angular distance, ω is the radian frequency or angular velocity.

We could obtain an equation for time by taking the first derivative of the phase. It would take the form

$$\frac{d\theta}{d\omega} = t \qquad\qquad \text{Equation 3-2}$$

In a similar manner, we can write an equation for phase for an electrical network. It can be shown that the time delay of a sinusoidal signal of frequency ω is determined by taking the first derivative of the minus phase equation in respect to the frequency ω.

$$t_{(\omega)} = \frac{d\,[-\,\theta_{(\omega)}]}{d\omega}$$

Time Delay Calculation

Let us now work an example: Given the transfer function $H(j\omega)$, determine the time delay $t_{(\omega)}$.

$$\frac{E_o}{E_i} = H(j\omega) = \frac{4\,j\omega + 3}{-\omega^2 + 9\,j\omega + 7}$$

Determining the phase equation,

$$\theta(\omega) = \arctan \frac{4\omega}{3} - \arctan \frac{9\omega}{7 - \omega^2}$$

The negative of the phase constant is:

$$\theta(\omega) = -\arctan \frac{4\omega}{3} + \arctan \frac{9\omega}{7 - \omega^2}$$

The negative of the phase constant is:

$$\theta_{(\omega)} = -\arctan \frac{4\omega}{3} + \arctan \frac{9\omega}{7 - \omega^2}$$

In evaluation of $t_{(\omega)}$, we have

$$t_{(\omega)} = \frac{d\left[-\theta_{(\omega)}\right]}{d\omega}$$

$$t_{(w)} = \frac{d}{d\omega}[-\arctan \frac{4\omega}{3}] + \frac{d}{d\omega}[\arctan \frac{9\omega}{7 - \omega^2}]$$

To evaluate $t_{(\omega)}$, we should use the following formula from calculus:

$$\frac{d}{dx} \arctan \frac{u}{v} = \frac{1}{1 + \left(\frac{u}{v}\right)^2} \frac{d}{dx}\left(\frac{u}{v}\right)$$

Let us now evaluate each part of the formula for $t_{(\omega)}$

$$\frac{d}{d\omega}[-\arctan \frac{4\omega}{3}] = -\frac{1}{1 + \left(\frac{4\omega}{3}\right)^2} \frac{d}{d\omega}\left(\frac{4\omega}{3}\right)$$

$$= -\frac{1}{\dfrac{9 + 16\omega^2}{9}} \frac{d}{d\omega}\left(\frac{4\omega}{3}\right)$$

$$= -\frac{9}{9 + 16\omega^2}\left(\frac{4}{3}\right)$$

$$= -\frac{12}{9 + 16\omega^2}$$

$$\frac{d}{d\omega}[\arctan \frac{9\omega}{7 - \omega^2}] = \frac{1}{1 + \left(\frac{9\omega}{7 - \omega^2}\right)^2} \frac{d}{d\omega}\left|\frac{9\omega}{7 - \omega^2}\right|$$

Evaluating, we obtain:

$$\frac{d}{d\omega} \arctan \frac{9\omega}{7 - \omega^2} = \frac{(7 - \omega^2)\dfrac{d\,9\omega}{d\omega} - 9\omega\dfrac{d}{d\omega}(7 - \omega^2)}{[7 - \omega^2]^2}$$

$$= \frac{(7-\omega^2)\,(9) - 9\,\omega\,[0 - 2\omega]}{[7-\omega^2]^2}$$

$$= \frac{63 - 9\,\omega^2 + 18\,\omega^2}{[\,7 - \omega^2\,]^2}$$

$$\frac{d}{d\omega}\arctan\frac{9\,\omega}{7-\omega^2} = \frac{1}{1 + \left(\dfrac{9\,\omega}{7-\omega^2}\right)^2}\left(\frac{63 - 9\,\omega^2 + 18\,\omega^2}{[\,7 - \omega^2\,]^2}\right)$$

$$= \frac{7 - \omega^{2\,2}}{[\,7 - \omega^2\,]^2 + 9\,\omega^2)}\ \frac{63 - 9\,\omega^2 + 18\,\omega^2}{[\,7 - \omega^2\,]^2}$$

$$= \frac{63 + 9\omega^2}{49 - 14\omega^2 + \omega^4 + 81\omega^2}$$

$$= \frac{63 + 9\,\omega^2}{\omega^4 + 67\,\omega^2 + 49}$$

The complete time equation then becomes:

$$t_{(\omega)} = \frac{-12}{9 + 16\,\omega^2} + \frac{63 + 9\,\omega^2}{\omega^4 + 67\omega^2 + 49}$$

$$= \frac{-12\,(\omega^4 + 67\omega^2 + 49) + (9 + 16\omega^2)\,(63 + 9\,\omega^2)}{(9 + 16\omega^2)\,(\omega^4 + 67\omega^2 + 49)}$$

$$= \frac{-12\,\omega^4 - 804\,\omega^2 - 588 + 567 + 81\omega^2 + 1008\omega^2 + 144\,\omega^2}{9\omega^4 + 603\,\omega^2 + 441 + 16\,\omega^6 + 1072\,\omega^4 + 784\,\omega^2}$$

$$= \frac{132\,\omega^4 + 285\,\omega^2 - 21}{16\,\omega^6 + 1089\omega^4 + 1387\,\omega^2 + 441}$$

At a frequency of $\omega = 1$ radian/second, we have a time delay of

$$t_{(1)} = \frac{132\,(1)^4 + 285\,(1)^2 - 21}{16\,(1)^6 + 1089\,(1)^4 + 1387\,(1)^2 + 441}$$

$$= \frac{396}{2933} = .1351\ \text{sec.}$$

Example 3-6. Given the circuit in Fig. 3-6, determine the time delay at 7 radians per second.

1. Write the mesh equations for the network and determine $H_{(s)}$:

$$E_i = I_1\,(SL_1 + I_1\,(R_1) - I_2\,(R_1)$$

$$0 = -I_1\,(R_1) + I_2\,(R_1 + SL_2 + R_2)$$

Since $L_1 = L_2 = L$, and $R_1 = R_2 = R$, the mesh equations become

$$E_i = I_1 (SL + R) - I_2 R$$

$$0 = -I_1 R + I_2 (2R + SL)$$

$$\Delta = \begin{vmatrix} SL + R\, , -R \\ -R,\, 2R + SL \end{vmatrix}$$

$$= (SL + R)(2R + SL) - (-R)(-R)$$

$$= 2SLR + S^2 L^2 + 2R^2 + SLR - R^2$$

$$= S^2 L^2 + 3 SLR + R^2$$

$$E_o = (I_2)R$$

$$I_2 = \frac{\Delta I_2}{\Delta}$$

$$I_2 = \frac{\begin{vmatrix} SL + R,\, E_i \\ -R\, ,\, 0 \end{vmatrix}}{\Delta}$$

$$I_2 = \frac{-(R)(E_i)}{\Delta}$$

$$I_2 = \frac{R E_i}{S^2 L^2 + 3SLR + R^2}$$

$$E_o = \frac{R E_i}{S^2 L^2 + 3SLR + R^2} R$$

$$\frac{E_o}{E_i} = \frac{R^2}{S^2 L^2 + 3SLR + R^2}$$

Since $S = j\omega$,

$$\frac{E_o}{E_i} = \frac{(10)^2}{-\omega^2 (2)^2 + 3 j\omega (2)(10) + (10)^2}$$

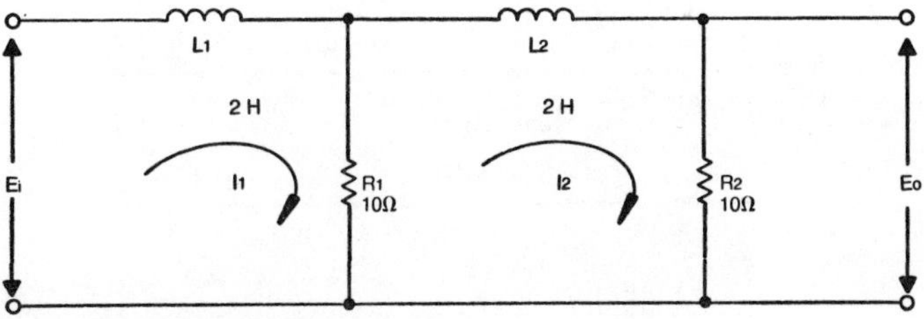

Fig. 3-6. Circuit for example 3-6.

71

$$= \frac{100}{-4\omega^2 + 60\,j\omega + 100}$$

$$= \frac{100}{100 - 4\omega^2 + 60\,j\omega}$$

$$\theta_{(\omega)} = -\arctan\frac{60\,\omega}{100 - 4\,\omega^2}$$

Finding $-\theta_{(\omega)}$, we have,

$$-\theta_{(\omega)} = \arctan\frac{60\,\omega}{100 - 4\,\omega^2}$$

If we now determine $t_{(\omega)}$, we have,

$$t_{(\omega)} = \frac{d}{d\omega}\left[\arctan\frac{60\,\omega}{100 - 4\,\omega^2}\right] = \left(\frac{1}{1 + \left[\frac{60\,\omega}{100 - 4\omega^2}\right]^2}\right)\left(\frac{d}{d\omega}\frac{60\,\omega}{100 - 4\,\omega^2}\right)$$

$$\frac{d}{d\omega}\frac{60\,\omega}{100 - 4\,\omega^2} = \frac{(100 - 4\omega^2)\dfrac{d\,60\omega}{d\,\omega} - 60\,\omega\dfrac{d\,(100 - 4\,\omega^2)}{d\,\omega}}{[\,100 - 4\,\omega^2\,]^2}$$

$$= \frac{(100 - 4\,\omega^2)\,(60) - 60\,\omega\,[\,0 - 8\omega\,]}{[\,100 - 4\,\omega^2\,]^2}$$

$$= \frac{6000 - 240\,\omega^2 + 480\,\omega^2}{[\,100 - 4\,\omega^2\,]^2}$$

$$= \frac{6000 + 240\,\omega^2}{[\,100 - 4\,\omega^2\,]^2}$$

Finding the equation for $t_{(\omega)}$, we have

$$t_{(\omega)} = \frac{1}{1 + \left[\frac{60\,\omega}{100 - 4\omega^2}\right]^2}\left[\frac{6000 + 240\omega^2}{[\,(100 - 4\,\omega^2)\,]}\right]$$

$$= \frac{[\,100 - 4\omega^2\,]^2}{[\,100 - 4\,\omega^2\,]^2 + (60\,\omega)^2}\left[\frac{6000 + 240\,\omega^2}{[\,100 - 4\,\omega^2\,]^2}\right]$$

$$= \frac{6000 + 240\,\omega^2}{10{,}000 - 800\,\omega^2 + 16\,\omega^4 + 3600\,\omega^2}$$

$$= \frac{6000 + 240\,\omega^2}{10{,}000 + 2800\omega^2 + 16\,\omega^4}$$

72

At $\omega = 7$ radians per second, we obtain

$$t_{(\omega)} = \frac{6000 + 240\ (7)^2}{10,000 + 2800\ (7)^2 + 16\ (7)^4}$$

$$= \frac{17,760}{185,616} = 0.954 \text{ sec.}$$

This simply means that a sinewave with a frequency of seven radians per second will be delayed by 0.0954 seconds from its entry at the input terminals of the network until it appears at the output terminals.

DELTA TO Y CONVERSIONS

One very powerful tool in deriving transfer functions is the delta-to-"Y" conversion formulas. In Fig. 3-7, we see the "Y" network, and its counterpart, the delta network. At times it may be easier to work with a "Y" network instead of a delta network, and vice versa.

To convert from the delta to the "Y" circuit, we use these formulas:

$$Z_a = \frac{Z_1 Z_2}{Z_1 + Z_2 + Z_3} \qquad \text{Equation 3-3}$$

$$Z_b = \frac{Z_2 Z_3}{Z_1 + Z_2 + Z_3} \qquad \text{Equation 3-4}$$

$$Z_c = \frac{Z_1 Z_3}{Z_1 + Z_2 + Z_3} \qquad \text{Equation 3-5}$$

To go from the "Y" to the delta circuit these formulas are used:

$$Z_1 = \frac{Z_a Z_c + Z_a Z_b + Z_b Z_c}{Z_b} \qquad \text{Equation 3-6}$$

$$Z_2 = \frac{Z_a Z_c + Z_a Z_b + Z_b Z_c}{Z_c} \qquad \text{Equation 3-7}$$

$$Z_3 = \frac{Z_a Z_c + Z_a Z_b + Z_b Z_c}{Z_a} \qquad \text{Equation 3-8}$$

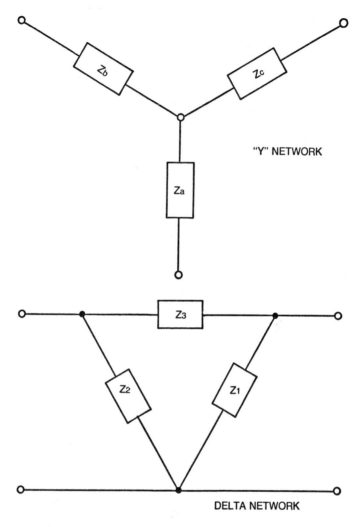

Fig. 3-7. Typical "Y" and Delta circuits.

Let us try an example with these equations.

Example 3-7. Convert the circuit in Fig. 3-8 to a "Y" network. We note that $Z_1 = 10$ ohms, $Z_2 = 20$ ohms, and $Z_3 = 30$ ohms.

$$\Delta = Z_1 + Z_2 + Z_3$$
$$= 10 + 20 + 30 = 60\,\Omega$$
$$Z_a = \frac{Z_1 Z_2}{\Delta} = \frac{(10)\,(20)}{60} = 3.33 \text{ ohms}$$

$$Z_b = \frac{Z_2 Z_3}{\Delta} = \frac{(20)(30)}{60} = 10 \text{ ohms.}$$

$$Z_c = \frac{Z_1 Z_3}{Z_1 + Z_2 + Z_3} = \frac{(10)(30)}{60} = 5 \text{ ohms.}$$

The resultant y network is shown in Fig. 3-9.

Example 3-8. Convert the "Y" circuit shown in Fig. 3-10 to a delta circuit.

We note that $Z_a = 10$ ohms, $Z_b = 20$ ohms, and $Z_c = 30$ ohms.

$$Y = Z_a Z_c + Z_a Z_b + Z_b Z_c$$
$$= (10)(30) + (10)(20) + (20)(30)$$
$$= 1100^2$$

$$Z_1 = \frac{\Delta'}{Z_b} = \frac{1100}{20} = 55 \text{ ohms.}$$

$$Z_2 = \frac{\Delta'}{Z_c} = \frac{1100}{30} = 36.7 \text{ ohms.}$$

$$Z_3 = \frac{\Delta'}{Z_a} = \frac{1100}{10} = 110 \text{ ohms.}$$

The equivalent delta circuit is shown in Fig. 3-11.

NULL NETWORK TRANSFER FUNCTION DERIVATIONS

One of the simplest methods of evaluating the transfer function for the bridge-tee network or twin-tee network is to use a Y (Wye) to Δ (Delta) conversion.

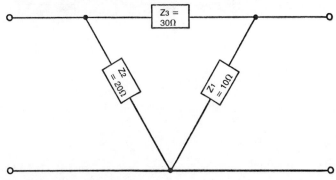

Fig. 3-8. Circuit for example 3-7.

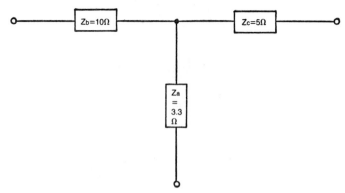

Fig. 3-9. Answer to example 3-7.

Bridge Tee

Figure 3-12 shows the bridge-tee network. The network can be simplified by redrawing it as shown in Fig. 3-13. Using the concept of Y to Δ conversion, we may calculate the impedances of the delta network as follows:

$$Z_3 = \frac{Z_a Z_b + Z_a Z_c + Z_b Z_c}{Z_a}$$

$$= \frac{R\left(\frac{1}{SC}\right) + R\left(\frac{1}{SC}\right) + \left(\frac{1}{SC}\right)\left(\frac{1}{SC}\right)}{R}$$

$$= \frac{2}{SC} + \left(\frac{1}{SC}\right)^2 \frac{1}{R}$$

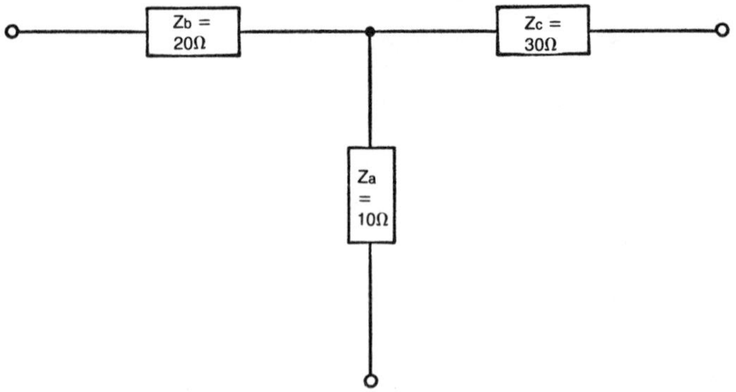

Fig. 3-10. Circuit for example 3-8.

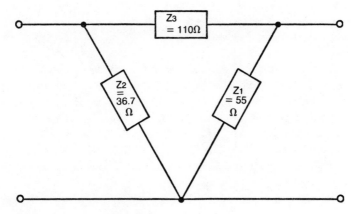

Fig. 3-11. Answer to example 3-8.

$$Z_1 = Z_2 = \frac{Z_a Z_b + Z_a Z_c + Z_b Z_c}{Z_c}$$

Because $Z_b = Z_c \dfrac{R\left(\dfrac{1}{SC}\right) + R\left(\dfrac{1}{SC}\right) + \left(\dfrac{1}{SC}\right)\left(\dfrac{1}{SC}\right)}{\dfrac{1}{SC}}$

$$= 2R + \frac{1}{SC}$$

We can now evaluate the transfer function by using the voltage divider theorem:

$$\frac{E_o}{E_i} = \frac{Z_1}{R \ 11 \ Z_3 + Z_1}$$

The resistor R, and Z3 in parallel give

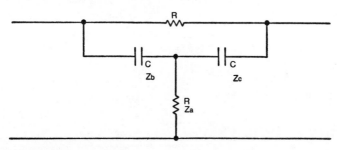

Fig. 3-12. Bridge-Tee network.

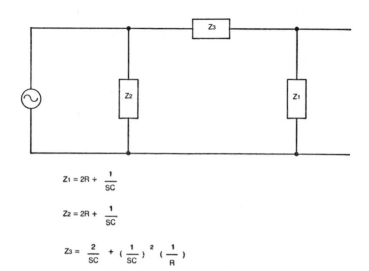

$$Z_1 = 2R + \frac{1}{SC}$$

$$Z_2 = 2R + \frac{1}{SC}$$

$$Z_3 = \frac{2}{SC} + \left(\frac{1}{SC}\right)^2 \left(\frac{1}{R}\right)$$

Fig. 3-13. Bridge-Tee network equivalent.

$$Z_3' = \frac{R(Z_3)}{R + Z_3}$$

$$Z_3' = \frac{\left[R\,\frac{2}{SC} + \left(\frac{1}{SC}\right)^2 \frac{1}{R} \right]}{R + \frac{2}{SC} + \left(\frac{1}{SC}\right)^2 \frac{1}{R}}$$

$$= \frac{\frac{2R}{SC} + \frac{1^2}{SC}}{R + \frac{2}{SC} + \frac{1^2}{SC}\,\frac{1}{R}}$$

Multiplying numerator and denominator by $(SC)^2 R$ we obtain

$$Z_3' = \frac{2R^2\,S\,C + R}{(SC)^2\,R^2 + 2\,S\,C\,R + 1}$$

We can then write the transfer function

$$\frac{E_o}{E_i} = \frac{Z_1}{Z_3' + Z_1} = \frac{2R + \dfrac{1}{SC}}{\dfrac{2R^2\,SC + R}{(SCR)^2 + 2SCR + 1} + 2R + \dfrac{1}{SC}}$$

If we multiply numerator and denominator by SC, we obtain

$$\frac{E_o}{E_i} = \frac{2\,RSC + 1}{\dfrac{2\,R^2\,(SC)^2 + SCR}{(SCR)^2 + 2SCR + 1} + 2RSC + 1}$$

We can then multiply numerator and denominator by $(SCR)^2 +$ $2SCR + 1$:

$$\frac{E_o}{E_i} = \frac{(2RSC + 1)\,[\,(SCR)^2 + 2SCR + 1\,]}{2R^2\,(SC)^2 + SCR +[\,2RSC + 1\,][\,(SCR)^2 + 2SCR + 1\,]}$$

The transfer function may be simplified by factoring $2RSC + 1$ out of the denominator:

$$\frac{E_o}{E_i} = \frac{(2RSC + 1)\,[\,(SCR)^2 + 2SCR + 1\,]}{(2\,RSC + 1)\,[\,RSC + (SCR)^2 + 2SCR + 1\,]}$$

$$\frac{E_o}{E_i} = \frac{(SCR)^2 + 2\,SCR + 1}{(SCR)^2 + 3SCR + 1}$$

By examination of the transfer function, you will find that there is no frequency which can cause the numerator to completely disappear. If you plot the magnitude versus frequency, you will note that the magnitude reaches a minimum at a frequency of:

$$f_n = \frac{1}{2\,\pi\,RC}$$

This is called the null frequency. The magnitude of the function is found by first substitution of $j\omega$ for S in the transfer function. The result is:

$$H\,(j\omega) \;=\; \frac{1 - (\omega CR)^2 + 2j\,\omega CR}{1 - (\omega CR)^2 + 3\,j\,\omega CR}$$

$$|\,H\,(j\omega)\,| = \frac{\sqrt{(1 - (\omega CR)^2\,)^2 + (2\omega CR)^2}}{\sqrt{(1 - (\omega CR)^2\,)^2 + (3\omega CR)^2}}$$

$$= \frac{\sqrt{1 - 2\,(\omega CR)^2 + (\omega CR)^4 + 4\,(\omega CR)^2}}{\sqrt{1 - 2\,(\omega CR)^2 + (\omega CR)^4 + 9\,(\omega CR)^2}}$$

$$= \frac{\sqrt{1 + 2\,(\omega CR)^2 + (\omega CR)^4}}{\sqrt{1 + 7\,(\omega CR)^2 + (\omega CR)^4}}$$

At the null frequency, the magnitude becomes

$$|H(j\omega)| = \frac{\sqrt{1 + 2\left(\frac{1}{CR}CR\right)^2 + \left(\frac{1}{CR}CR\right)^4}}{\sqrt{1 + 7\left(\frac{1}{CR}CR^2\right) + \left(\frac{1}{CR}CR\right)^4}}$$

$$= \frac{\sqrt{1 + 2 + 1}}{\sqrt{1 + 7 + 1}}$$

$$= \frac{\sqrt{4}}{\sqrt{9}}$$

$$= \frac{2}{3}$$

Twin Tee

Like the bridge-tee network, the twin-tee network also reduces very easily to a wye-to-delta conversion. Figure 3-14 shows the twin-tee network. We can simplify the network by redrawing it as shown in Fig. 3-15. Using similar conversion formulas as before, we obtain the following results:

$$Z_1 = Z_2 = \frac{Z_a Z_b + Z_a Z_c + Z_b Z_c}{Z_b}$$

$$= \frac{R\left(\frac{1}{2SC}\right) + R\left(\frac{1}{2SC}\right) + R \times R}{R}$$

$$= \frac{1}{CS} + R$$

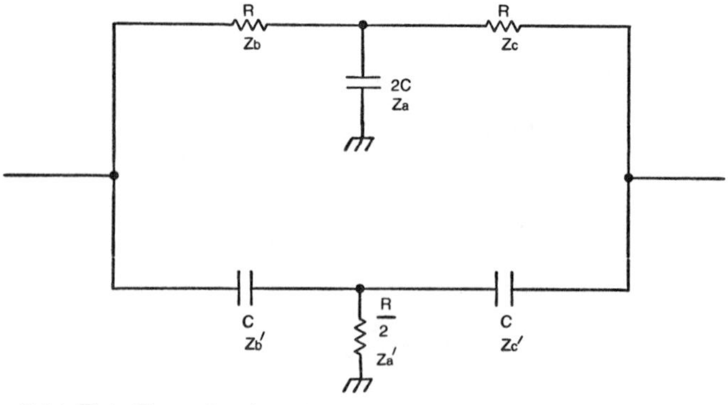

Fig. 3-14. Twin-Tee network.

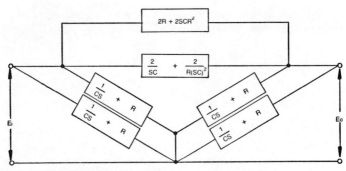

Fig. 3-15. Twin Tee network equivalent.

$$Z_3 = \frac{Z_a Z_b + Z_a Z_c + Z_b Z_c}{Z_a}$$

$$= \frac{R\left(\frac{1}{2SC}\right) + R\left(\frac{1}{2SC}\right) + R \times R}{\frac{1}{2SC}} = R + R + 2SCR^2$$

$$= 2R + 2SCR^2$$

$$Z_1' = Z_2' = \frac{Z_a' Z_b' + Z_a' Z_c' + Z_b' Z_c'}{Z_b'}$$

$$= \frac{\left(\frac{1}{CS}\right)\left(\frac{R}{2}\right) + \left(\frac{1}{CS}\right)\left(\frac{R}{2}\right) + \left(\frac{1}{CS}\right)\left(\frac{1}{CS}\right)}{\frac{1}{CS}} = R + \frac{1}{CS}$$

$$Z_3' = \frac{Z_a' Z_b' + Z_a' + Z_c' Z_b' Z_c'}{Z_a'}$$

$$= \frac{\left(\frac{1}{CS}\right)\left(\frac{R}{2}\right) + \left(\frac{1}{CS}\right)\left(\frac{R}{2}\right) + \left(\frac{1}{CS}\right)\left(\frac{1}{CS}\right)}{\frac{R}{2}}$$

$$= \frac{1}{CS} + \frac{1}{CS} + \frac{2}{R}\left(\frac{1}{CS}\right)^2$$

Since the Z_2 and Z_2' components appear across the generator, they do not affect the transfer function (assuming negligible source impedance and very high load impedance).

Figure 3-16 shows the end result. We must now determine the transfer function. We can write

$$\frac{E_o}{E_i} = \frac{Z_1 Z_1'}{Z_3 Z_3' Z_1 Z_1'}$$

The parallel combination of Z_1 and Z_1' gives

$$Z_1'' = \frac{\frac{1}{CS} + R}{2} = \frac{1}{2CS} + \frac{R}{2} = \frac{2 + 2RCS}{4}$$

The parallel combination of the Z_3 terms gives

$$Z_3'' = \frac{(2R + 2SCR^2)\left(\dfrac{2}{SC} + \dfrac{2}{R(SC)^2}\right)}{2R + 2SCR^2 + \dfrac{2}{SC} + \dfrac{2}{R(SC)^2}}$$

$$= \frac{\dfrac{4R}{SC} + \dfrac{4}{(SC)^2} + 4R^2 + \dfrac{4R}{SC}}{2R + 2SCR^2 + \dfrac{2}{SC} + \dfrac{2}{R(SC)^2}}$$

$$= \frac{4R^2SC + 4R + 4R^3(SC)^2 + 4R^2SC}{2R^2(SC)^2 + 2R^3S^3C^3 + 2RSC + 2}$$

The transfer function then becomes

$$\frac{E_o}{E_i} = \frac{\dfrac{2 + 2RCS}{4CS}}{\dfrac{2 + 2RCS}{4CS} + \dfrac{2R^2SC + 2R + 2R^3(SC)^2 + 2R^2SC}{R^2(SC)^2 + R^3S^3C^3 + RSC + 1}}$$

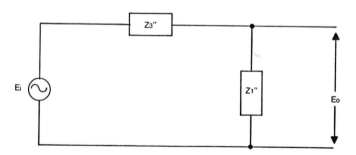

Fig. 3-16. Circuit to determine transfer function for twin tee network.

This may be simplified into

$$\frac{E_o}{E_i} = \frac{2\,S^4R^4C^4 + 4S^3\,R^3\,C^3 + 4S^2R^2C^2 + 4SCR + 2}{2\,S^4R^4C^4 + 12S^3R^3C^3 + 20S^2R^2C^2 + 12SRC + 2}$$

Dividing by two and factoring,

$$\frac{E_o}{E_i} = \frac{(S^2R^2C^2 + 1)\,(S^2R^2C^2 + 2SCR + 1)}{(S^2R^2C^2 + 2SRC + 1)\,(S^2R^2C^2 + 4SRC + 1)}$$

$$\frac{E_o}{E_i} = \frac{S^2R^2C^2 + 1}{S^2R^2C^2 + 4SCR + 1}$$

Dividing numerator and denominator by R^2C^2, we obtain:

$$\frac{E_o}{E_i} = \frac{S^2 + \left(\dfrac{1}{RC}\right)^2}{S^2 + S\left(\dfrac{4}{RC}\right) + \left(\dfrac{1}{RC}\right)^2}$$

In terms of ω we have,

$$\frac{E_o}{E_i} = \frac{-\omega^2 + \left(\dfrac{1}{RC}\right)^2}{-\omega^2 + \left(\dfrac{1}{RC}\right)^2 + j\,\omega\left(\dfrac{4}{RC}\right)}$$

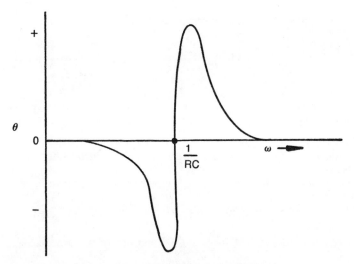

Fig. 3-17. Phase versus frequency for a twin-tee network.

83

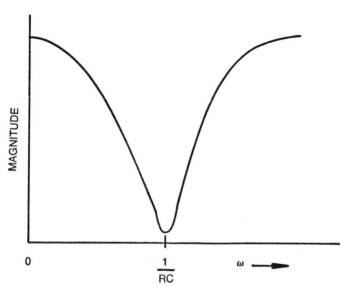

Fig. 3-18. Magnitude versus frequency for a twin-tee network.

The magnitude equation can be determined as

$$\frac{E_o}{E_i} = \frac{\sqrt{\left[\left(\frac{1}{RC}\right)^2 - \omega^2\right]^2}}{\sqrt{\left[\left(\frac{1}{RC}\right)^2 - \omega^2\right]^2 + \left[\frac{4\omega}{RC}\right]^2}}$$

If we evaluate the phase equation we obtain

$$\theta = \theta_n - \theta d = \arctan \frac{0}{\left(\frac{1}{RC}\right)^2 - \omega^2} - \arctan \frac{\dfrac{4\omega}{RC}}{\left(\frac{1}{RC}\right)^2 - \omega^2}$$

You may ask why we don't neglect the first term in the phase equation because there is a zero in the numerator of the fraction belonging to the numerator phase. The reason is that the denominator is not always a positive number. Since the denominator is $\left(\frac{1}{RC}\right)^2 - \omega^2$, there is a frequency where ω^2 is larger than $\left(\frac{1}{RC}\right)^2$. The numerator then becomes $\theta_n = \arctan \frac{0}{-a}$ where a is a number. The arctan of $\frac{0}{-a}$ is 180° and not 0° as is arctan $\frac{0}{a}$. Therefore, the phase starts out at a value of 0°, since at $\omega = 0$ the

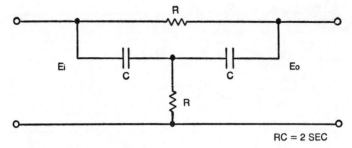

Fig. 3-19. Circuit for example 3-9.

phase is $\theta = \arctan \dfrac{0}{\left(\dfrac{1}{RC}\right)^2} - \arctan \dfrac{0}{\left(\dfrac{1}{RC}\right)^2} = 0°$

Then, as frequency rises, the first term (θ_n) remains at 0° but the second term starts to get larger in the negative direction. As frequency approaches a value of $\dfrac{1}{RC}$, the phase of the second term (θd) approaches $-90°$. When the frequency is just slightly smaller than $\dfrac{1}{RC}$ the phase is approximately $-90°$. As the frequency goes by $\dfrac{1}{RC}$, and gets to a value which is slightly larger than $\dfrac{1}{RC}$, the phase is

$$\theta = 180° - \arctan \dfrac{\dfrac{4\omega}{RC}}{\left(\dfrac{1}{RC}\right)^2 - \omega^2}$$

As frequency approaches infinity, the second term approaches 180°. Therefore the phase follows the plot shown in Fig. 3-17. In Chapter Two we discussed the band reject filter. A plot of mag-

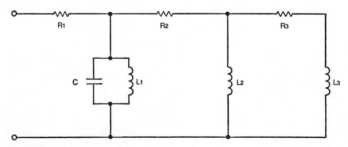

Fig. 3-20. Circuit for problems 1 and 4.

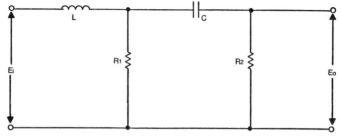

Fig. 3-21. Circuit for problem 5.

nitude for the circuit of Fig. 3-14 is shown in Fig. 3-18. This is an excellent circuit for use as a band-rejection filter since it has a very sharp null region. The bridge-tee network is not as sharp in response as the twin tee. The bridge tee normally has a very broad null with a smaller attenuation factor than the twin tee.

Example 3-9. Given the following transfer function, draw a circuit that can be represented by

$$H_{(s)} = \frac{4S^2 + 4S + 1}{4S^2 + 6S + 1}$$

Note that this equation is that of a bridge-tee network. Inspection shows that the value of RC must be 2 seconds. This is because

$$H_{(S)} = \frac{4S^2 + 4S + 1}{4S^2 + 6S + 1} = \frac{(SRC)^2 + 2SRC + 1}{(SRC)^2 + 3SRC + 1}$$

The null frequency is 0.5 radian per second. If R is 1 ohm, then C must be 2 farads. If R is 10, C must be 0.2 farads etc.

The circuit that the equation represents is shown in Fig. 3-19.

CHAPTER SUMMARY

In this chapter, we have derived many transfer-function equations. Regardless of how complicated a transfer function is, it has

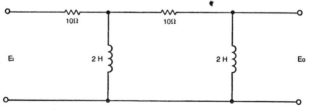

Fig. 3-22. Circuit for problem 6.

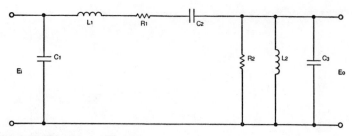

Fig. 3-23. Circuit for problem 9.

two parts: magnitude, $| H(j\omega) |$, and phase, θ. As the frequency changes the magnitude and phase vary.

The magnitude is the size of the ratio of output voltage to input voltage, while the phase is the angle by which the output voltage either leads or lags the input voltage.

From the phase equation, we can derive the time delay equation which depends on the first derivative of frequency.

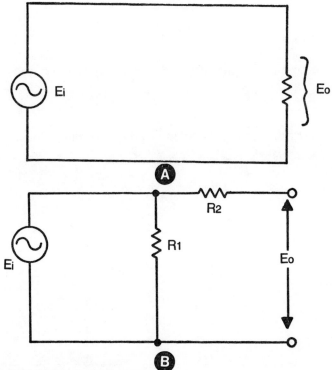

Fig. 3-24. Circuits for problem 10.

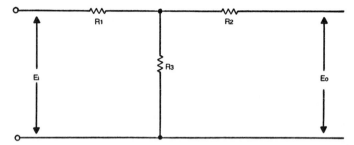

Fig. 3-25. Another circuit for problem 10.

Problems

1. Determine the transfer function for the circuit shown in Fig. 3-20.

2. (Research question). Extract a stable transfer function from the following equations:

a. $|H(j\omega)| = \dfrac{169 + \omega^2}{\sqrt{1 + \omega^6}}$

b. $|H(j\omega)| = \dfrac{54\,\omega}{\sqrt{1 + \omega^2}}$

3. Compute the time delay equation for this transfer function

$$H_{(s)} = \dfrac{4S + 13}{S^2 + 9s + 7}$$

4. For the network shown in Fig. 3-20, find the following:

a. magnitude equation
b. phase equation
c. time delay equation

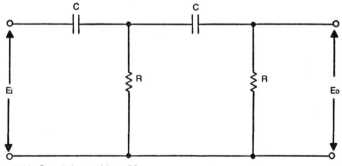

Fig. 3-26. Circuit for problem 11.

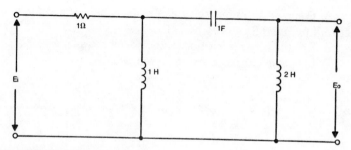

Fig. 3-27. Circuit for problem 12.

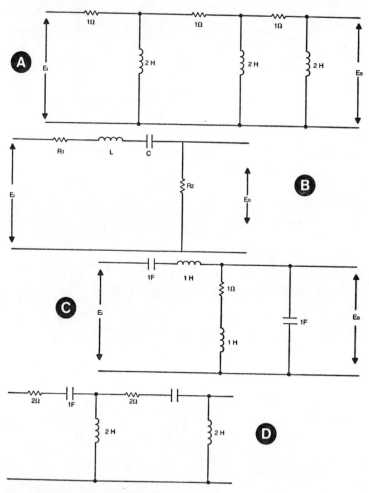

Fig. 3-28. Circuits for problem 13.

5. Compute the time delay at **7** radians per second for the network in Fig. 3-21.

6. Find the transfer function for the circuit shown in Fig. 3-22.

7. Draw the circuits that can be represented by:

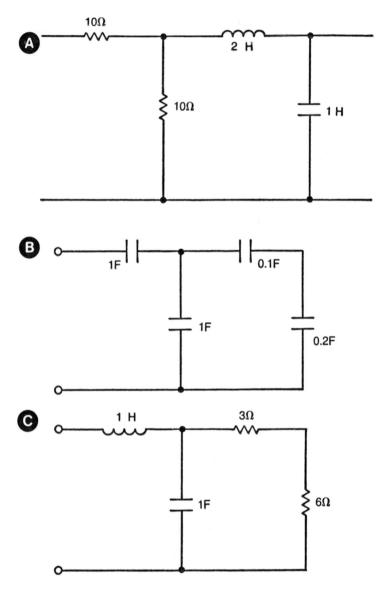

Fig. 3-29. More circuits for problem 13.

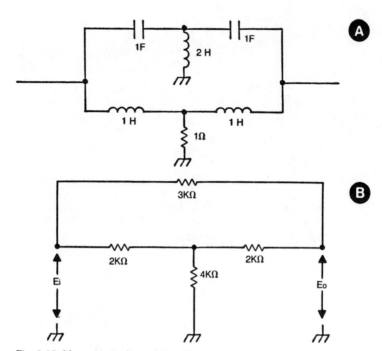

Fig. 3-30. More circuits for problem 13.

a. $H_{(s)} = \dfrac{.01S^2 + .2S + 1}{.01S^2 + .35 + 1}$

b. $H_{(s)} = \dfrac{S^2 + 40000}{S^2 + 800S + 40,000}$

8. (Research question). Draw a circuit that can be represented by

$$H_{(s)} = \dfrac{S^4 + 200\ S^2 + 10,000}{S^4 + 80S^3 + 1800S^2 + 8000S + 10000}$$

9. Find the transfer function for the circuit shown in Fig. 3-23.

10. Find the transfer function for the circuits shown in Figs. 3-24 and 3-25.

11. Find the transfer function of the circuit in Fig. 3-26.

12. Find the transfer function for the circuit shown in Fig. 3-27 and plot magnitude and phase versus frequency.

13. Plot magnitude and phase versus frequency for the circuits shown in Fig. 3-28.

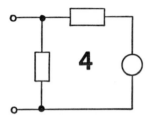

Impedance and Frequency Scaling

Some of the circuits we have synthesized so far may have seemed to have rather unrealistic values for their components. This is because many times it is easier to synthesize a model with convenient values, and then scale its frequency response either up or down to obtain the desired response.

FREQUENCY SCALING

For example, the filter shown in Fig. 4-1 has a cutoff frequency of 1 radian per second. The values used in building the filter are abnormally large for the capacitors, large for the inductors, and rather small for the resistors. First of all, why does the circuit have a cutoff of 1 radian / second? To see this we must write the mesh equations:

$$E_i = I_1\left(R_1 + SL_1 + \frac{1}{CS}\right) - I_2\left(\frac{1}{CS}\right) \qquad \text{Equation 4-1}$$

$$0 = -I_1\left(\frac{1}{CS}\right) + I_2\left(\frac{1}{CS} + SL_2 + R_2\right) \qquad \text{Equation 4-2}$$

To determine the transfer function requires an evaluation of the system determinant, known as Δ :

$$\Delta = \left|R_1 + SL_1 + \frac{1}{CS}, -\frac{1}{CS} - \frac{1}{CS}, \frac{1}{CS} + SL_2 + R_2\right|$$

$$= \left(R_1 + SL_1 + \frac{1}{CS}\right)\left(\frac{1}{CS} + SL_2 + R_2\right) - \left(\frac{-1}{CS}\right)\left(\frac{-1}{CS}\right)$$

$$=\frac{R_1}{CS}+SL_2R_1+R_1R_2+\frac{SL_1}{CS}+S^2L_1L_2+R_{2s}L_1+\left(\frac{1}{CS}\right)^2\frac{SL_2}{CS}+\frac{R_2}{CS}-\left(\frac{-1}{CS}\right)^2$$

$$=\frac{R_1}{CS}+SL_2R_1+SL_1R_2+\frac{L^1S}{CS}+\frac{L^2S}{CS}+R_1R_2+S^2L_1L_2+\frac{R_2}{CS}$$

Also, we must determine I_2. First, find ΔI_2.

$$\Delta I_2 = \left|R_1 + SL_1 + \frac{1}{CS}\ ,\ E_i\ -\frac{1}{CS}\ ,\ 0\right|$$

$$= (-)\ E_i\ \left(\frac{-1}{CS}\right)=\frac{E_i}{CS}$$

I_2 is the ratio of ΔI_2 to Δ. This is given by

$$I_2 = \frac{\Delta I_2}{\Delta} \qquad \text{Equation 4-3}$$

$$=\frac{\dfrac{E_i}{CS}}{S^2 L_1 L_2 + SL_2R_1 + SL_1R_2 + \dfrac{R_1}{CS}+\dfrac{R_2}{CS}+\dfrac{L_1}{C}+\dfrac{L_2}{C}+R_1R_2}$$

Multiplication of numerator and denominator by CS gives

$$I_2 = \frac{E_i}{S^3CL_1L_2 + S^2L_2R_1C + S^2L_1R_2C + R_1 + R_2 + L_1S + R_1R_2CS}$$

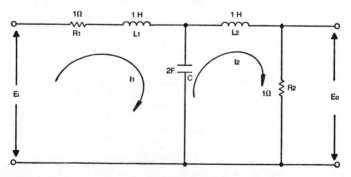

Fig. 4-1. Filter with a cutoff frequency of one radian per second.

The output voltage is given by

$$E_o = I_2 (R_2) \qquad \text{Equation 4-4}$$

Equation 4-4a

$$E_o = \frac{E_i R_2}{S^3 C L_1 L_2 + S^2 L_2 R_1 C + S^2 L_1 R_2 C + R_1 + R_2 + L_1 S + L_2 S + R_1 R_2 CS}$$

The transfer function then becomes

$$\frac{E_o}{E_1} = \frac{R_2}{S^2 C L_1 L_2 + S^2 L_2 R_1 C + S^2 L_1 R_2 C + R_1 + R_2 + \left(\begin{smallmatrix}\text{line is}\\\text{con'd}\\\text{below}\end{smallmatrix}\right)}$$

$$\overline{L_1 S + L_2 S + R_1 R_s CS}$$

Substituting values,

$$\frac{E_o}{E_i} = \frac{1}{S^3 (2)(1)(1) + S^2 (1)(1)(2) + S^2 (1)(1)(2) + \left(\begin{smallmatrix}\text{line is}\\\text{con'd}\\\text{below}\end{smallmatrix}\right)}$$

$$\overline{1 + 1 + 1 (S) + 1 (S) + (1)(1)(2) S}$$

$$= \frac{1}{-j2 \omega^3 - 4 \omega^2 + 4 \omega + 2}$$

The magnitude of the transfer function is

$$H (j\omega) = \frac{1}{-j2\omega^3 - 4\omega^2 + 4j\omega + 2}$$

$$|H (j\omega)| = \frac{1}{\sqrt{(2-4\omega^2)^2 + (4\omega - 2\omega^3)^2}}$$

$$= \frac{1}{\sqrt{4 - 16\omega^2 + 16\omega^4 + 16\omega^2 - 16\omega^4 + 4\omega^6}}$$

$$= \frac{1}{\sqrt{4\omega^6 + 4}}$$

At $\omega = 0$, $|H(j\omega)| = \frac{1}{2}$

94

The frequency at which the magnitude reaches .707 of the magnitude at zero frequency is found by setting the magnitude equal to $\frac{1}{\sqrt{2}}$ (H (jω) at $\omega = 0$).

$$\left(\frac{1}{2}\right) \frac{1}{\sqrt{2}} = \frac{1}{\sqrt{4\omega^6 + 4}}$$

taking reciprocals,

$$2\sqrt{2} = \sqrt{4\omega^6 + 4}$$

Squaring both sides,

$$4 \ (2) = 4\omega^6 + 4$$

$$8 = 4\omega^6 + 4$$

$$4 = 4\omega^6$$

$$\omega^6 = 1$$

$$\omega = 1 \text{ rad/sec}$$

Suppose we would like to have a cutoff frequency of 4 rad/sec, what must we do? Obviously, we must change some values. We know that resistance is not a function of frequency, so the value of resistance should remain the same. We know that resonant frequency of a simple LC network is inversely proportional to the square root of LC. If inductance and capacitance go down by a factor of four, frequency goes up by a factor of four. In other words, if an inductance of L and a capacitance of C gives a resonant frequency of F_o, then a value of inductance $\frac{L}{n}$ and a capacitance of $\frac{C}{n}$ gives a resonant frequency of

$$\frac{1}{\sqrt{\frac{L}{n} \times \frac{C}{n}}} = \frac{n}{\sqrt{LC}} \text{ , or } nw_o.$$

The original transfer function becomes that given by Equation 4-5.

$$\frac{E_o}{E_i} = \frac{1}{S^3 \left(\frac{2}{4}\right)\left(\frac{1}{4}\right)\left(\frac{1}{4}\right) + S^2 \left(\frac{1}{4}\right)(1)\left(\frac{2}{4}\right) + S^2 \left(\frac{1}{4}\right)(1)\left(\frac{2}{4}\right) + 1} \ \text{(line is cont'd)}$$

$$+ 1 + \frac{1}{4}S + \frac{1}{4}S + (1)(1)\left(\frac{2}{4}S\right)$$

$$= \frac{1}{\dfrac{S^3}{32} + \dfrac{S^2}{8} + \dfrac{S^2}{8} + 2 + \dfrac{2S}{4} + \dfrac{2S}{4}}$$

$$= \frac{1}{\dfrac{S^3}{32} + \dfrac{S^2}{4} + S + 2}$$

$$\frac{E_o}{E_i} = \frac{32}{S^3 + 8S^2 + 32S + 64} = \frac{32}{-j\omega^3 - 8\omega^2 + 32j\omega + 64}$$

$$= \frac{32}{64 - 8\omega^2 + 32j\omega - j\omega^3} \qquad \text{Equation 4-5}$$

Evaluating the magnitude equation,

$$|H(j\omega)| = \frac{32}{\sqrt{64 - 8\omega^2) + (32\omega - \omega^2)^2}}$$

$$= \frac{32}{\sqrt{4096 - 1024\omega^2 + 64\omega^2 + 1024\omega^2 - 64\omega^4 + \omega^6}}$$

$$= \frac{32}{\sqrt{\omega^6 + 4096}}$$

Suppose $\omega = 0$,

$$|H(j\omega)| = \frac{32}{\sqrt{4096}} = \frac{32}{64} = \tfrac{1}{2}$$

Suppose $\omega = 4$

$$|H(j\omega)| = \frac{32}{\sqrt{(4)^6 + 4096}} = \frac{32}{\sqrt{4096 + 4096}}$$

$$= \frac{32}{\sqrt{2(4096)}} = \frac{1}{\sqrt{2}} \times \frac{32}{\sqrt{4096}}$$

$$= \frac{1}{\sqrt{2}} \times \frac{32}{64} = \left(\frac{1}{\sqrt{2}}\right)\tfrac{1}{2}$$

As you can see, the response is down to $\frac{1}{\sqrt{2}}$ ($\frac{1}{2}$) at a frequency of 4 radians/second. Thus the cutoff frequency has been enlarged four times.

At another frequency, such as 2 radians/second, what would be the response for the circuit with a cutoff of one radian/second, and for the circuit with a cutoff of 4 radian/second? For a cutoff of 1 radian/second, the transfer function was given by

$$|H(j\omega)| = \frac{1}{\sqrt{4\omega^6 + 4}}$$

At W = 2 radian/second,

$$|H(j\omega)| = \frac{1}{\sqrt{4(2)^6 + 4}} = \frac{1}{\sqrt{4(64) + 4}}$$

$$|H(j\omega)| = \frac{1}{\sqrt{260}} \cong .063$$

For a cutoff of 4 radian/seconds,

$$= \frac{32}{\sqrt{\omega^6 + 4096}} = \frac{32}{\sqrt{(2)^6 + 4096}}$$

$$= \frac{32}{\sqrt{64 + 4096}} = \frac{32}{\sqrt{4160}} = 0.496$$

This shows that the transfer function has the same value at the cutoff frequencies but it is different at other frequencies. Figure 4-2 is a plot of the functions for the filter with the three different cutoff frequencies.

Example 4-2. What would be the transfer function if the cutoff frequency is to be 10 rad/sec? Solution: Since frequency is to be ten times as large, then inductance and capacitance must be divided by ten. Using Equation 4-4a, use $R_1 = 1$ ohm, $R_2 = 1$ ohm, $L_1 = L_2 = \frac{1}{10}$ H, and $C = \frac{2}{10}$ F.

$$\frac{E_o}{E_i} = \frac{1}{S^3\left(\frac{2}{10}\right)\left(\frac{1}{10}\right)\left(\frac{1}{10}\right) + S^2\left(\frac{1}{10}\right)(1)\left(\frac{2}{10}\right) + S^2\left(\frac{1}{10}\right)(1)\left(\frac{2}{10}\right)} \quad \text{(line is cont'd below)}$$

$$+ 1 + 1 + \frac{1\,S}{10} + \frac{1\,S}{10} + (1)\,(1)\left(\frac{2}{10}\right)$$

$$= \frac{1}{\dfrac{2S^3}{1000} + \dfrac{4S^2}{100} + 1 + 1 + \dfrac{2\,S}{10} + \dfrac{2\,S}{10}}$$

$$= \frac{1000}{2S^3 + 40S^2 + 1000 + 1000 + 200S + 200\,S}$$

$$= \frac{1000}{2S^3 + 40S^2 + 400\,S + 2000}$$

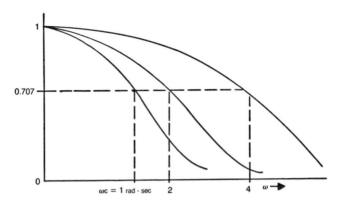

Fig. 4-2. Behavior of filter in Fig. 4-1 with three different cutoff frequencies.

$$\frac{E_o}{E_i} = \frac{1000}{-j2\,\omega^3 - 40\,\omega^2 + 400\,j\,\omega + 2000}$$

$$= \frac{1000}{2000 - 40\,\omega^2 + 400\,j\,\omega - j2\,\omega^3}$$

$$= \frac{1000}{\sqrt{(2000 - 40\,\omega^2)^2 (400\,\omega - 2\,\omega^3)^2}}$$

$$= \frac{1000}{\sqrt{4 \times 10^6 - 16 \times 10^4\,\omega^2 + 1600\,\omega^4 + 16 \times 10^4\omega^2 - 1600\,\omega^4 + 4\,\omega^6}}$$

$$= \frac{1000}{\sqrt{4\,\omega^6 + 4 + 10^6}}$$

Check if $\omega = 0$:

$$\frac{E_o}{E_i} = \frac{1000}{\sqrt{4 \times 10^6}} = \frac{1}{2}$$

If $\omega = 10$:

$$\frac{E_o}{E_i} = \frac{1000}{\sqrt{4 \times (10)^6 + 4 \times 10^6}} \qquad = \frac{1000}{\sqrt{8 \times 10^6}}$$

$$= \frac{1000}{\sqrt{2\,2 \times 10^3}} \qquad = \frac{1}{2} \times \frac{1}{\sqrt{2}}$$

In the previous problems, we have seen what happens when we scale the response of a circuit. Table 4-1 also illustrates this. Note that altering a capacitance or inductance value by $\frac{1}{n}$ results in the cutoff frequency changing to n times the original value.

However, what happens at other frequencies? Suppose we have a transfer function as given by

$$|H(j\omega)| = \frac{1}{\sqrt{4\omega^6 + 4}} \qquad \text{Equation 4-6}$$

If we were to use the transfer function scaled to a cutoff frequency of 4 radian/second we get

$$|H(j\omega)| = \frac{32}{\sqrt{\omega^6 + 4096}}$$

At 0 radian/second we get

$$= \frac{32}{\sqrt{(0)^6 + 4096}} = \tfrac{1}{2}$$

At 2 radian/seconds we get

$$|H(j\omega)| = \frac{32}{\sqrt{(2)^6 + 4096}} \cong \tfrac{1}{2}$$

Table 4-1. Frequency-Scaling Formulas.

Old (Original)	New
R	R
C	$\dfrac{C}{\eta}$
L	$\dfrac{L}{\eta}$
$\eta =$	$\dfrac{\text{new frequency}}{\text{old frequency}}$

At 4 radian/seconds we get

$$| H (j\omega)| = \frac{32}{\sqrt{(4)^6 + 4096}} = 0.355$$

At a cutoff frequency of 1 radian/second, the function becomes

$$| H (j\omega)| = \frac{1}{\sqrt{4(1)^6 + 4}} = \frac{1}{\sqrt{8}} = \frac{1}{2} \times \frac{1}{\sqrt{2}}$$

If the cutoff frequency was 1 radian/second and the operating frequency were the same.

Suppose we examine the behavior of the circuit at 2 radian/seconds. The response is

$$| H (j\omega)| = \frac{1}{\sqrt{4 (2)^6 + 4}} = .062$$

This is a much smaller volume than that of the filter with a cutoff of 4 radian/seconds. When the circuit is scaled to a new cutoff frequency of 4 radian/seconds, the response again will be .062 when the frequency was twice the cutoff, or eight radians.

Suppose we had evaluated the circuit with a cutoff of 1 radian/second at 10 radian/seconds. The response would be

$$| H (j\omega)| = \frac{1}{\sqrt{4 (10)^6 + 4}} = \frac{1}{2 \times 10^3} = 0.5 \times 10^{-3}$$

Suppose a frequency of eight radians per second is put into the expression. The result will be

$$| H (j\omega)| = \frac{32}{\sqrt{(8)^6 + 4096}} = \frac{32}{\sqrt{65536 + 4096}}$$

$$= \frac{32}{\sqrt{69,632}} = \frac{32}{263} = .062$$

Note that the response at twice the cutoff frequency for each filter was the same.

In evaluating the circuit with a cutoff of 4 radian/seconds at 10 radian/seconds we would have

$$| H (j\omega)| = \frac{32}{\sqrt{(4 \times 10)^6 + 4096}}$$

$$= \frac{32}{\sqrt{(4096) (10)^6 + 4096}}$$

$$= \frac{32}{64 \times 10^3} = .5 \times 10^{-3}$$

Example 4-3. Figure 4-3 shows an LC parallel circuit. The inductance is 1 mH and the capacitance if 40 μF. Find the resonant frequency and compute the values of L and C which will give the circuit a resonant frequency of three times higher than, and one third of, the original value.

$$F_r = \frac{1}{2\pi\sqrt{LC}} = \frac{1}{2\pi\sqrt{1 \times 10^{-3})(4 \times 10^{-5})}}$$

$$= \frac{.159}{2 \times 10^{-4}} = .0795 \times 10^4 = 795 \text{ Hz}$$

To obtain a resonant frequency of three times F_r, or 3 (795) = 2385 Hz, inductance and capacitance values should be one third the original values, or

$$L_{new} = \frac{L}{3} = \frac{1 \text{ mH}}{3} = .333 \text{ mH}$$

$$C_{new} = \frac{C}{3} = \frac{4\,0\,\mu F}{3} = 13.3 \ \mu F$$

To obtain a resonant frequency of one third F_r, or $\frac{795}{3} = 265$ Hz, inductance and capacitance values should be three times the original value, or

$$L_{new} = 3L = 3(1mH) = 3 \text{ mH}$$

$$C_{new} = 3C = 3 (40 \text{ F}) = 120 \ \mu \text{ F}.$$

Example 4-4. Given the circuit in Fig. 4-4A, determine the values of inductance and resistance to result in doubling this circuit's cutoff frequency. Also calculate the new cutoff frequency.

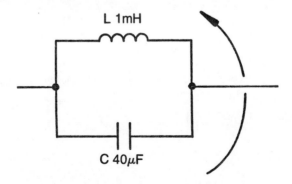

L 1mH

C 40μF

Fig. 4-3. Circuit for example 4-3.

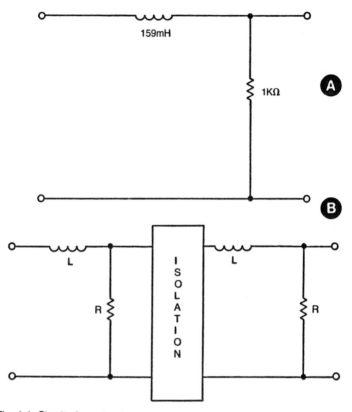

Fig. 4-4. Circuits for example 4-4.

From a previous problem, we know that the cutoff frequency of one section is given by

$$f_c = \frac{R}{2 \pi L}$$

$$= \frac{1000}{2 \pi (.159)} = 1000 \text{ Hz}$$

For the isolated sections, Fig. 4-4B, the resultant cutoff frequency is found by the formula

$$f_{c_2} = f_c \sqrt{2^{\left(\frac{1}{n}\right)} - 1} \bigg| \quad (n=2)$$

$$= 1000 \sqrt{2^{\frac{1}{2}} - 1}$$

$$= 1000 \sqrt{1.414 - 1}$$

$$= 1000\sqrt{0.414}$$

$$= 1000\,(.64)$$

$$= 640\text{ Hz}$$

To double the cutoff frequency, we must divide all inductors by 2. Resistors will remain the same.

$$R_{new} = R_{old} = 1000\,\Omega$$
$$L_{new} = \frac{L_{old}}{2} = \frac{.159}{2} = .0795\text{H}$$

The new cutoff frequency for the entire filter will be twice the original cutoff frequency, or 2 (640) = 1280 Hz.

Where does the formula for the cutoff frequency of several stages come from?
Using the formula for one section's magnitude, we have

$$|\,H\,(j\omega)| = \frac{1}{\sqrt{1 + \left|\dfrac{\omega}{\omega_c}\right|^2}}$$
Equation 4-7

The response of n sections would be given by

$$|H(j\omega)|^n = \left[\frac{1}{\sqrt{1 + \left|\dfrac{\omega}{\omega_c}\right|^2}}\right]^n$$
Equation 4-8

The overall cutoff frequency is found by setting

$$|\,H(j\omega)|^n = \frac{1}{\sqrt{2}} \text{ and solving for } \omega.$$

$$\frac{1}{\sqrt{2}} = \left[\frac{1}{\sqrt{1 + \left(\dfrac{\omega}{\omega_c}\right)^2}}\right]^n$$

$$\frac{1}{\sqrt{2}} = \frac{1}{\left[\sqrt{1 + \left(\dfrac{\omega}{\omega_c}\right)^2}\right]^n}$$

$$\sqrt{2} = \left[\sqrt{1 + \left(\dfrac{\omega}{\omega_c}\right)^2}\right]^n$$

$$\left(\sqrt{2}\right)^2 = \left[\sqrt{1 + \left(\dfrac{\omega}{\omega_c}\right)^2}\right]^{2n}$$

$$2 = \left[1 + \left(\frac{\omega}{\omega_c}\right)^2\right]^n$$

$$2^{\frac{1}{n}} = \left[1 + \left(\frac{\omega}{\omega_c}\right)^2\right]^{\frac{n}{n}}$$

$$2^{\frac{1}{n}} = 1 + \left(\frac{\omega}{\omega_c}\right)^2$$

$$2^{\frac{1}{n}} - 1 = \left(\frac{\omega}{\omega_c}\right)^2$$

$$\omega^2 = (\omega_c)^2 (2^{\frac{1}{n}} - 1)$$

$$\omega = \omega_c (2^{\frac{1}{n}} - 1) \qquad \text{Equation 4-9}$$

As shown in Equation 4-9, the overall frequency of stages depends on the integer n. The larger the value of n the smaller is the value of overall cutoff frequency. Table 4-2 illustrates the results of the formula.

IMPEDANCE SCALING

A network may be scaled in impedance as well as frequency, as shown, for example, in the circuit of Fig. 4-5. At the resonant frequency, the circuit has an impedance of 10 ohms. Suppose we wanted the impedance at resonance to be 100 ohms. We then simply make the resistor 100 ohms. However, suppose we wanted the impedance of the circuit to be always ten times its original value. In other words, if the original circuit had an impedance of 50 ohms at 200 Hz the new circuit would have an impedance of 500 ohms at 200 Hz; if the original circuit had an impedance of 20 ohms at 400 Hz the new circuit would have an impedance of 200 ohms at 400 Hz, etc. How is this done? Look at the simple circuit of Fig. 4-6. The impedance is given by

$$Z = j\omega L + R + \frac{1}{j\omega C} \qquad \text{Equation 4-10}$$

If you would like to change the impedance to some new value (K_z) (Z) where K_z is the impedance scaling constant, simply multiply $j\omega L$ by K_z, R by K_z, and $\frac{1}{j\omega C}$ by K_z:

$$K_z Z = j\omega L (K_z) + R (K_z) + \left(\frac{1}{j\omega C}\right) K_z \qquad \text{Equation 4-11}$$

Table 4-2. Bandwidth Reduction Factors.

Number of Stages (Identical)	Bandwidth
1	f_c
2	0.64 f_c
3	0.51 f_c
4	0.43 f_c

Since the frequency must remain the same, the new values for components for impedance scaling would be given as follows:

$$L_{new} = L\,(K_z)$$
$$R_{new} = R\,(K_z)$$
$$C_{new} = \frac{C}{K_z}$$

Example 4-5. Scale the following circuit in impedance so that its impedance is twenty times as large as it is now. See Fig. 4-7 for the circuit.

1. Make $K_z = 20$
2. L_{new} $= L\,(20)$
              $= 3\text{ mH }(20)$
              $= 60\text{ mH}$
    R_{new} $= (20)\,R$
              $= 20\,(40)$
              $= 800\text{ ohms}$
    $C_{new} = \dfrac{50\ \mu\text{F}}{20} = 2.5\ \mu\text{F}.$

Example 4-6. Scale the following circuit in impedance so that the impedance is one tenth as large as it is now. See Fig. 4-8A for the circuit.

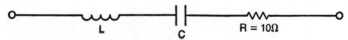

Fig. 4-5. Basic R-L-C series network.

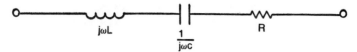

Fig. 4-6. Basic series-circuit component.

1. $\quad$ Make $K_z \quad = \dfrac{1}{10}$

2. $\quad L_{new} \quad = \dfrac{L}{10}$

$\quad\quad\quad\quad\quad\quad = \dfrac{40 \text{ mH}}{10}$

$\quad\quad\quad\quad\quad\quad = 4 \text{ mH}$

$\quad\quad R_{new} \quad = 50 \, (K_z)$

$\quad\quad\quad\quad\quad\quad = 50\left(\dfrac{1}{10}\right)$

$\quad\quad\quad\quad\quad\quad = 5 \text{ ohms}$

$\quad\quad C_{new} \quad = \dfrac{30 \ \mu\text{F}}{K_z}$

$\quad\quad\quad\quad\quad\quad = \dfrac{30 \ \mu\text{F}}{\dfrac{1}{10}}$

$\quad\quad\quad\quad\quad\quad = 300 \ \mu\text{F}.$

Example 4-7. Scale the following circuit in impedance so that the impedance is forty times as large as it is now. See Fig. 4-9 for details. We know that the equation for impedance must be

$$Z = \frac{R}{1 + j\omega CR}$$

```
o——uuu——| |——w——o
   3mH    50μF    40Ω
```

Lnew = 60mH
Cnew = 2.5μF
Rnew = 800Ω

Fig. 4-7. Circuit for example 4-5.

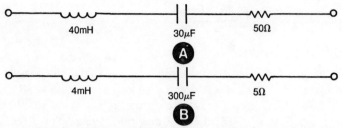

Fig. 4-8. Original circuit for example 4-6, (A) and impedance-scaled circuit (B).

The new value of R must be

$$R_{new} = R (K_z)$$
$$= 80 (40)$$
$$= 3200 \text{ ohms}$$

The new value of C must be

$$C_{new} = \frac{C}{K_z}$$

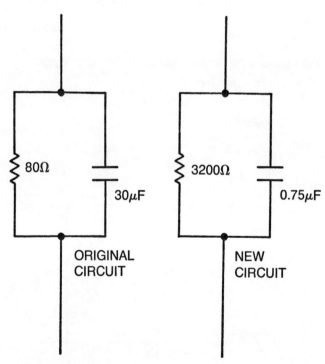

ORIGINAL CIRCUIT

NEW CIRCUIT

Fig. 4-9. Circuits for example 4-7.

$$= \frac{30 \ \mu F}{40}$$

$$= 0.75 \ \mu F$$

Example 4-8. Scale the circuit in Fig. 4-10 so that the new cutoff frequency is 2 radians per second. Use a capacitor value of 50 μ F.

First of all, the value of the capacitor as it is now is 10 μF; the new capacitor is to be 50 μF. The new cutoff frequency is 2 radians per second and the original cutoff frequency is 1000 radian/seconds. We know that for impedance scaling, we have

$$C_{new} = \frac{C}{K_z}$$

For frequency scaling, we have

$$C_{new} = \frac{C}{n}$$

For both frequency and impedance scaling the new value of C would be given by

$$C_{new} = \frac{C}{(K_z) \ (n)} \qquad \text{Equation 4-10}$$

In this example, the scaling constant for frequency must be

$$n = \frac{\text{new cutoff frequency}}{\text{old cutoff frequency}}$$

$$= \frac{2 \ \text{rad/sec}}{1000 \ \text{rad/sec}}$$

$$= 2 \ \times \ 10^{-3}$$

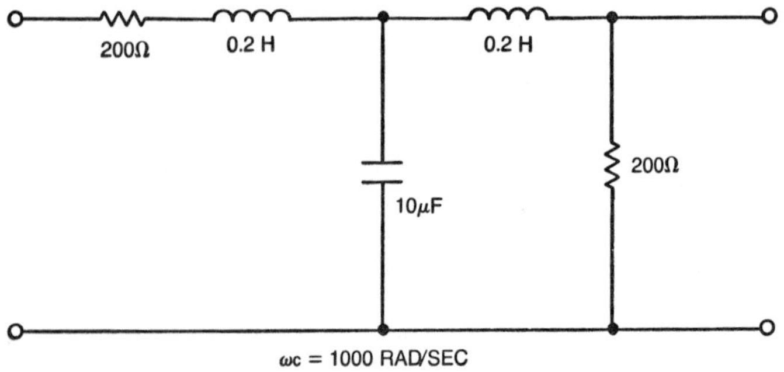

Fig. 4-10. Circuit for example 4-8.

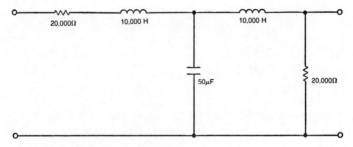

Fig. 4-11. Scaled circuit for example 4-8.

We must solve for the impedance scaling constant:

$$C_{new} = \frac{C}{(K_z)\,(n)}$$

$$50\ \mu\text{F} = \frac{10\ \mu\text{F}}{(K_z)\,(2 \times 10^{-3})}$$

$$K_z = \frac{10\ \mu\text{F}}{C_{new}\,(2 \times 10^{-3})}$$

$$= \frac{10 \times 10^{-6}}{50 \times 10^{-6} \times 2 \times 10^{-3}}$$

$$= \frac{1}{10^{-2}}$$

$$= 100$$

The new value of resistances must be given by

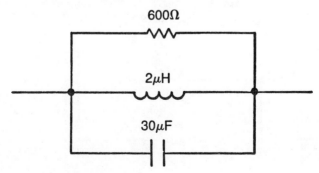

Fig. 4-12. Circuit for problem 1.

$$R_{new} = K_z R$$
$$= (100)\ (200)$$
$$= 20\ k\Omega$$

The new value of inductance must be

$$L_{new} = K_z\ (L)$$

for impedance scaling, and for frequency scaling it must be

$$L_{new} = \frac{L}{n}$$

$$L_{new} = \frac{L\ K_z}{n}$$

$$= \frac{0.2\ H\ (100)}{2 \times 10^{-3}}$$

$$= 10,000H.$$

Figure 4-11 shows the resultant circuit after scaling.

CHAPTER SUMMARY

This chapter has covered the methods by which networks are scaled in frequency as well as impedance. Remember that, when scaling by frequency, the original capacitor and inductor value are divided by the scaling constant, while no change occurs in the resistor value.

When scaling by impedance, the original values of resistance and inductance are multiplied by the scaling constant, while capacitance is divided by the scaling constant.

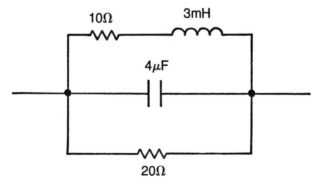

10Ω   3mH

4μF

20Ω

Fig. 4-13. Circuit for problem 2.

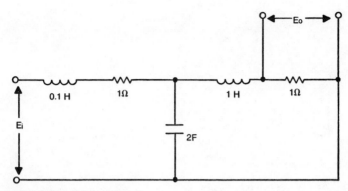

Fig. 4-14. Circuit for problem 3.

Problems

1. Scale the circuit of Fig. 4-12 so that the new resonant frequency is 8000 Hz.

2. Scale the circuit of Fig. 4-13 so that the impedance at any frequency would be one half the impedance of the circuit at the frequency in Fig. 4-13.

3. Scale the circuit of Fig. 4-14 so that the new cutoff frequency is 40,000 hertz, and the capacitor value is $1 \ \mu F$. The circuit as shown has a cutoff frequency of 1 radian/second.

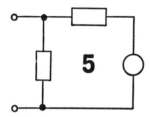

Butterworth and Chebyshev Low-Pass Filters

There is, at times, a need for a filter which approaches the magnitude shape of the ideal low-pass filter. Two very important filters that follow this shape are the Butterworth and Chebyshev low-pass filters.

BUTTERWORTH LOW-PASS FILTERS

A realizable magnitude response $H(j\omega)$ which approximates an ideal low pass filter is given by

$$|H(j\omega)| = \frac{1}{\sqrt{1 + f (f^2)^n}} \qquad \text{Equation 5-1}$$

The part $f (f^2)^n$ is a function of frequency. However, note that it is related to a power of frequency squared. In other words, the function may be related to frequency squared, frequency to the fourth power, as well as frequency to the sixth power, etc.

A function suitable for $f (f^2)^n$ happens to be $(\omega)^{2n}$ where $n = 1, 2, 3, 4$ etc. This then makes the magnitude response as given by

$$|H(j\omega)| = \frac{1}{\sqrt{1 + \omega^{2n}}} \qquad \text{Equation 5-2}$$

This equation is for the n_{th} order Butterworth with a cutoff of one radian per second, or 0.159 Hz.

As frequency becomes very large, the magnitude response becomes that given by

$$|H(j\omega)| = \frac{1}{\sqrt{\omega^{2n}}} = \frac{1}{\omega^n} \qquad \text{Equation 5-3}$$

The dB response in relation to frequency is given

$$AdB(\omega) = 20 \log |H(j\omega)| \qquad \text{Equation 5-4}$$

For frequencies much larger than one radian per second, the response becomes

$$AdB(\omega) = 20 \log \frac{1}{\omega^n} \qquad \text{Equation 5-5}$$

This reduces to that given by

$$AdB(\omega) = -20 \log \omega^n \qquad \text{Equation 5-6}$$

The formulas given so far have been for a cutoff of 1 radian/second.

For cutoff frequencies other than 1 radian per second, each formula must be modified by replacing ω by $\frac{\omega}{\omega c}$ where ωc is the cutoff frequency.

For example, the magnitude response of the Butterworth filter becomes

$$|H(j\omega)| = \frac{1}{\sqrt{1 + \left(\frac{\omega}{\omega c}\right)^{2n}}} \qquad \text{Equation 5-7}$$

Normally, we must work with an expression which depends on the square of the magnitude response, which is

$$|H(j\omega)|^2 = \frac{1}{1 + \left(\frac{\omega}{\omega c}\right)^{2n}} \qquad \text{Equation 5-8}$$

For simplicity, in our analysis we will deal with Butterworth filters with a cutoff frequency of 1 radian/second. This will make the calculations easier. After the calculations are complete, the filter can be scaled for a higher or lower cutoff frequency.

ORDER OF A BUTTERWORTH FILTER

The order of a Butterworth filter determines the rate at which the filter response drops off. Figure 5-1 illustrates a typical Butterworth filter magnitude of order one, and cutoff at one radian per second. Note that the response remains near unity until reaching the cutoff frequency. At the cutoff, the response has dropped to 0.707 of the response at near zero frequency.

Figure 5-2 shows the dB response, which starts at zero and drops to -3dB at the cutoff frequency. Above the cutoff frequency, the response drops by about n (20 dB) per decade, where n is the filter order.

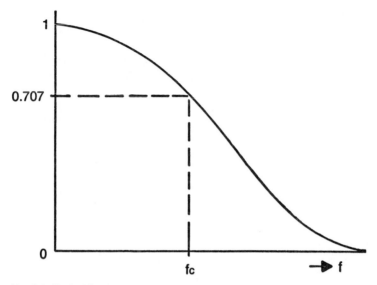

Fig. 5-1. Typical Butterworth filter response.

The response can be estimated by the approximate magnitude given by Equation 5-6.

At a frequency of 1 radian/second, the response is given by the following:

$$AdB_{(\omega)} = -20 \log \omega^n \Big|_{\substack{\omega=1 \\ n=2}}$$

$$= -20 \log \omega^2$$

$$= -20 \log (1)^2$$

$$= -20 \log 1 = 0dB$$

At a frequency of 10 rad/sec, we have

$$AdB_{(\omega)} = -20 \log \omega^n \Big|_{\substack{\omega=10 \\ n=2}}$$

$$= -20 \log \omega^2$$

$$= -20 \log (10)^2$$

$$= -20 \log (100)$$

$$= -40 \text{ dB}$$

At a frequency of 100 rad/sec, we have

$$AdB_{(\omega)} = -20 \log (100)^2$$
$$= -20 \log 10,000$$
$$= -80 \text{ dB}$$

And, at 1000 rad/sec, we have

$$AdB_{(\omega)} = -20 \log (1000)^2$$
$$= -20 \log (1000000)$$
$$= -120 \text{ dB}$$

A plot of this filter response is given in Fig. 5-3. For a filter of the second order, the response falls by 40 dB per decade beyond the cutoff frequency.

Likewise, for a sixth order Butterworth filter, the response would fall by n (20) = 6 (20), or 120 dB per decade.

Extraction of a Stable Transfer Function from the Desired Magnitude Response

Suppose we were given the magnitude response as follows:

$$H(j\omega) = \frac{1}{\sqrt{1 + (\omega)^{2n}}} \qquad \text{Equation 5-9}$$

First of all we would square the function, obtaining

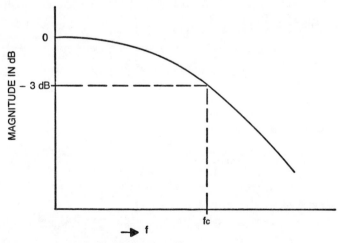

Fig. 5-2. Butterworth response in terms of dB.

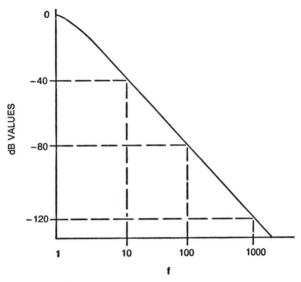

Fig. 5-3. Plot of $-20\omega^2$ versus frequency.

$$|H(j\omega)|^2 = \frac{1}{1+(s)^{2n}}$$ Equation 5-10

Then, remembering that $S=j\omega$ and $s^2=-\omega^2$, we write

$$\left|H(s)^2\right| = \frac{1}{\frac{1}{2}+(-s^2)^n}$$ Equation 5-11

We realize that $H(s)$ is made of two functions, namely $H(s)$ which is stable, and $H(-s)$ which is unstable. We can then write

$$\left|H(s)^2\right| = H(s)\,H(-s)$$ Equation 5-12

Parts $H(s)$ and $H(-s)$ can be thought of as functions $\frac{N(s)}{D(s)}$ and $\frac{N(-s)}{D(-s)}$ respectively. Where $D(s)$ is the denominator of $H(s)$, and $D(-s)$ is the denominator of $H(-s)$. $N(s)$ and $N(-s)$ are the respective numerators.

It is then obvious that we can now state

$$H(s)\,H(-s) = \frac{N(s)\,N(-s)}{D(s)\,D(-s)}$$ Equation 5-13

The product of $D(s)$ and $D(-s)$ must be equal to the denominator of $H(s)^2$, which is

$$D(s)\,D(-s) = 1 + (-s^2)^n$$ Equation 5-14

Example 5-1. Find a stable transfer function from

116

$$|H(j\omega)| = \frac{1}{\sqrt{1+\omega^2}}$$

First, square $|H(j\omega)|$, obtaining $|H(j\omega)|^2$:

$$|H(j\omega)|^2 = \frac{1}{1+\omega^2}$$

Substitute $S = j\omega$.

$$H(s)^2 = \frac{1}{1+(-S^2)} = \frac{1}{1-S^2}$$

Factor the denominator, obtaining

$$H(s)^2 = \frac{1}{(1-S)(1+S)}$$

We can now state that $D(-s) = 1-s$ and $D(s) = 1+s$. The unstable part is $1-s$ and the stable part is $1+s$. We can then write the stable transfer function:

$$H(s) = \frac{1}{1+S} \text{ , or } H(j\omega) = \frac{1}{1+j\omega}$$

Example 5-2. Find a stable transfer function for

$$|H(j\omega)| = \frac{1}{\sqrt{1+\omega^4}}$$

By inspection, we know that this is a second order filter, since $\omega^4 = \omega^{2n}$ where n is the order which must be 2. If $n = 2$, $D(s)D(-s) = 1 + (-S^2)^n$, which is $D(s)D(-s) = 1 + (s^2)^2 = 1 + S^4$. The quantity $1 + s^4$ may be factored by completing the square as follows:

$$1+S^4 = [S^4 + 2S^2 + 1] - 2S^2$$

$$= (S^2 + 1)^2 - (\sqrt{2}\,S)^2$$

$$= (S^2 + \sqrt{2}S + 1)(S^2 - \sqrt{2}S + 1)$$

The unstable part is $S^2 - \sqrt{2}S + 1$, while the stable part is $S^2 + \sqrt{2}S + 1$. The stable transfer function is then

$$H(s) = \frac{1}{S^2 + \sqrt{2}\,S + 1}$$

The previous two examples were really not very difficult because the order was very small. However, consider the Butterworth filter with a transfer function of

$$|H(j\omega)| = \frac{1}{\sqrt{1+\omega^6}}$$

By inspection, $\omega^6 = \omega^{2n}$ where n = 3. This indicates a third-order Butterworth filter.

Squaring the magnitude response, we obtain

$$|H(j\omega)|^2 = \frac{1}{1+\omega^6}$$

The denominator must be D(s) D(−s) = 1− S^6. This equation must be factored, which is not very simple. To solve the equation for its roots would dictate its factors. We may then write

$$1 + (-S^2)^n = (S-S_1)(S-S_2)(S-3)(\ldots(S-S_{2n-1})).$$

The roots of the equation are S_1, S_2 $\cdots$ up through 2n−1. To solve the equation $1+(-S^2)^n$:

$$S_i = -\sin\frac{(2_i-1)\pi}{2n} + j\cos\frac{(2_i-1)\pi}{2n} \qquad \text{Equation 5-15}$$

where n = 3 and i = 0, 1 . . . 2n−1, or 5.

Solving for the roots, we obtain

$$S_o = -\sin\frac{(2.0-1)\pi}{2(3)} + j\cos\frac{(2.0-1)\pi}{2(3)}$$

$$= -\sin\frac{(-\pi)}{6} + j\cos\frac{(-\pi)}{6}$$

$$= -\sin[-30°] + j\cos[-30°]$$

$$= -[-0.5] + j[0.866]$$

$$= 0.5 + j\,0.866$$

$$S_1 = -\sin\frac{(2.1-1)\pi}{2(3)} + j\cos\frac{(2.1-1)\pi}{2(3)}$$

$$= -\sin\frac{(\pi)}{6} + j\cos\frac{(\pi)}{6}$$

$$= -\sin(30°) + j\cos(30°)$$

$$= -(.5) + j0.866$$

$$= -0.5 + j0.866$$

$$S_2 = -\sin\frac{(2 \times 2 - 1)\,\pi}{2\,(3)} + j\cos\frac{(2 \times 2-1)\,\pi}{2\,(3)}$$

$$= -\sin\frac{(\pi)}{2} + j\cos\frac{(\pi)}{2}$$

$$= -\sin 90° + j\cos 90°$$

$$= -1 + j0$$

$$S_3 = -\sin\frac{(2\times3-1)\pi}{2\,(3)} + j\cos\frac{(2\times3-1)\,\pi}{2\,(3)}$$

$$= -\sin\frac{(5\,\pi)}{6} + j\cos\frac{(5\,\pi)}{6}$$

$$= -\sin (150°) + j\cos (150°)$$

$$= -0.5 + j\,(-0.866)$$

$$= -0.5 - j\,0.866$$

$$S_4 = -\sin\frac{(2\times4-1)\,\pi}{2\,(3)} + j\cos\frac{(2\times4-1)\,\pi}{2\,(3)}$$

$$= -\sin\frac{(7\,\pi)}{6} + j\cos\frac{(7\,\pi)}{6}$$

$$= -\sin (210°) + j\cos (210)°$$

$$= --.5 + j-.866$$

$$= 0.5 - j\,0.866$$

$$S_5 = -\sin\frac{(2\times5-1)\,\pi}{2\,(3)} + j\cos\frac{(2\times5-1)\,\pi}{2\,(3)}$$

$$= -\sin\frac{(9)\,\pi}{6} + j\cos\frac{(9)\,\pi}{6}$$

$$= -\sin 270° + j\cos 270°$$

$$= -(-1) + j\,0$$

Listing the roots, we have:

$$S_0 = 0.5 + j0.866$$

$$S_1 = -0.5 + j0.866$$

$$S_2 = -1 + j0$$

$$S_3 = -0.5 - j0.866$$
$$S_4 = 0.5 - j0.866$$
$$S_5 = 1 + j0$$

Half of the roots will lay on the left-hand section of the S plane, and half of the roots will lay on the right-hand section. We can plot these roots and then decide which to choose. Figure 5-4 shows the S plane with the respective roots. Roots S_0, S_4, and S_5 are on the right side while S_1, S_2, and S_3 are on the left side. Thus, S_1, S_2, and S_3 are our selection.

Now place these roots in the equation for the denominator of the transfer function to obtain

$$\begin{aligned} D_{(s)} &= (S-S_1)\ (S-S_2)\ (S-S_3) \qquad \text{Equation 5-16}\\ &= (S-[-0.5+j0.866]\ )\ (S-[-1])\ (S-\ [0.5-j0.866]\ )\\ &= (S+0.5-j0.866)\ (S+1)\ (S+0.5+j0.866) \end{aligned}$$

Multiplying:

$$S + 0.5 - j0.866$$
$$\underline{S + 0.5 + j0.866}$$

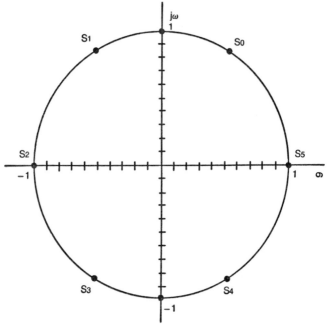

Fig. 5-4. The "S" plane.

120

$$S^2 + 0.5S - j0.866S$$
$$\quad + 0.5S + .25 - j0.433$$
$$\qquad + j0.866\ S + j0.433 + 0.75$$

$$
\begin{array}{l}
S^2 + S + 1 \\
S^2 + S + 1 \\
S\ \ + 1 \\
\hline
S^3 + S^2 + S \\
\quad\ S^2 + S + 1 \\
\hline
S^3 + 2S^2 + 2S + 1
\end{array}
$$

The answer is thus given by

$$H_{(s)} = \frac{1}{S^3 + 2S^2 + 2S + 1} \qquad \text{Equation 5-17}$$

As a further example, let us look at the 4th order Butterworth filter. We would like to find the transfer function for a filter with the magnitude equation given by

$$|H(j\omega)| = \frac{1}{\sqrt{1 + \omega^8}} \qquad \text{Equation 5-18}$$

By inspection, $\omega^8 = \omega^{2n}$, where $n = 4$.

This is how the fourth order Butterworth magnitude equation would appear, assuming a cutoff frequency of one radian per second. Squaring the magnitude response, we would obtain

$$|H(j\omega)^2| = \frac{1}{1 + \omega^8} \qquad \text{Equation 5-19}$$

The denominator must be $D(s)\ D(-s) = 1 + (-s^2)^n$, or $1+S^8$.

Solving the equation as before, we obtain the following square roots:

$$S_0 = -\sin\frac{(2\times0-1)\ \pi}{2\ (4)} \qquad + j\cos\frac{(2\times0-1)\ \pi}{2\ (4)}$$

$$\quad = -\sin\frac{(-\pi)}{8} \qquad + j\cos\frac{(-\pi)}{8}$$

$$\quad = 0.383 \quad + j\ 0.924$$

$$S_1 = -\sin\frac{(2\times1-1)\ \pi}{2\ (4)} \quad + j\cos\frac{(2\times1-1)\ \pi}{2\ (4)}$$

$$= -\sin \frac{(\pi)}{8} \qquad\qquad + j \cos \frac{(\pi)}{8}$$

$$= -0.383 \qquad\qquad + j\, 0.924$$

$$S_2 = -\sin \frac{(2\times2-1)\,\pi}{2\,(4)} \qquad + j \cos \frac{(2\times2-1)\,\pi}{2\,(4)}$$

$$= -\sin \frac{(3\,\pi)}{8} \qquad\qquad + j \cos \frac{(3\,\pi)}{8}$$

$$= -0.924 \qquad\qquad + j\, 0.383$$

$$S_3 = -\sin \frac{(2\times3-1)\,\pi}{2\,(4)} \qquad + j \cos \frac{(2\times3-1)\,\pi}{2\,(4)}$$

$$= -\sin \frac{(5\pi)}{8} \qquad\qquad + j \cos \frac{(5\pi)}{8}$$

$$= -0.924 \qquad\qquad - j\, 0.383$$

$$S_4 = -\sin \frac{(2\times4-1)\,\pi}{2\,(4)} \qquad + j \cos \frac{(2\times4-1)\,\pi}{2\,(4)}$$

$$= -\sin \frac{(7\pi)}{8} \qquad\qquad + j \cos \frac{(7\pi)}{8}$$

$$= -0.383 \qquad\qquad - j\, 0.924$$

$$S_5 = -\sin \frac{(2\times5-1)\,\pi}{2\,(4)} \qquad + j \cos \frac{(2\times5-1)\,\pi}{2\,(4)}$$

$$= -\sin \frac{(9\pi)}{8} \qquad\qquad + j \cos \frac{(9\pi)}{8}$$

$$= 0.383 \qquad\qquad - j\, 0.924$$

$$S_6 = -\sin \frac{(2\times6-1)\,\pi}{2\,(4)} \qquad + j \cos \frac{(2\times6-1)\,\pi}{2\,(4)}$$

$$= -\sin \frac{(11)\,\pi}{8} \qquad\qquad + j \cos \frac{(11)\,\pi}{8}$$

$$= 0.924 \qquad\qquad - j\, 0.383$$

$$S_7 = -\sin \frac{(2\times7-1)\,\pi}{2\,(4)} \qquad + j \cos \frac{(2\times7-1)\,\pi}{2\,(4)}$$

$$= \frac{-\sin (13\pi)}{8} \qquad + j \frac{\cos (13\pi)}{8}$$

$$= 0.924 \qquad\qquad + j\, 0.383$$

Now choose the stable roots. For those with little experience, it is wise to draw the S plane and pick the roots which are stable. Figure 5-5 shows the S plane and the associated roots. Those roots that are stable are:

$$S_1 = -0.383 \qquad + j0.924$$
$$S_2 = -0.924 \qquad + j0.383$$
$$S_3 = -0.924 \qquad + j0.383$$
$$S_4 = -0.383 \qquad - j0.924$$

The denominator will then be

$$D_{(s)} = (S-S_1)\,(S-S_2)\,(S-S_3)\,(S-S_4)$$

$$= (S - [-0.383 + j0.924\,])\,(S- [-0.924 + j0.383\,] \times (S - [-0.924 - j0.383])\,)\,(S-[-0.383 - j0.924])$$

$$= (S + 0.383 - j0.924)\,(S + 0.924 - j0.383) \times (S + 0.924 + j0.383)\,(S + 0.383 + j0.924)$$

Arranging the multiplication in terms of conjugate:

$$D_{(s)} = (S + 0.383 - j0.924)\,(S + 0.383 + j0.924) \times (S + 0.924 - j0.383)\,(S + 0.924 + j0.383)$$

$$= (S^2 + 0.766S + 1)\,(S^2 + 1.848S + 1)$$

$$= S^4 + 2.614S^3 + 3.416S^2 + 2.614S + 1$$

The transfer function then becomes the ratio of the numerator to denominator, which is given by

$$H(s) = \frac{1}{S^4 + 2.614\, S^3 + 3.416\, S^2 + 2.614\, S + 1} \qquad \text{Equation 5-18}$$

If you were asked to determine the phase equation you would simply gather the real and imaginary parts and proceed. For example, the transfer function as a function of $j\omega$ is

$$H(j\omega) = \frac{1}{(-\omega^2)^2 + 2.614\,(-j\omega^3) + 3.416\,(-\omega^2) + 2.614\,j\omega + 1}$$

$$= \frac{1}{\omega^4 - j2.614\,\omega^3 - 3.416\,\omega^2 + 2.614\,j\omega + 1}$$

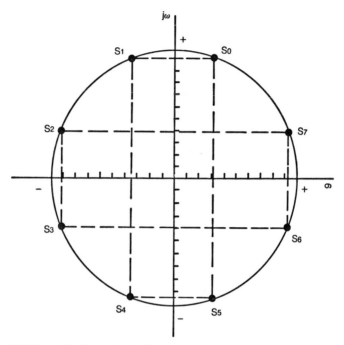

Fig. 5-5. Associated roots for $1+s^8$.

There is no imaginary part in the numerator so the phase is zero. There is a complex denominator which gives a phase of

$$\theta_d = \arctan \frac{2.614\,\omega - 2.614\,\omega^3}{\omega^4 - 3.416\,\omega^2 + 1}$$

The total phase is $\theta = \theta_n - \theta_d$, or

$$\theta = -\arctan \frac{2.614\,\omega - 2.614\,\omega^3}{\omega^4 - 3.416\,\omega^2 + 1}$$

Once the transfer function has been established, we can determine the manner in which the amplitude response and phase response behave.

CHEBYSHEV FILTER

Another special network exists which has a characteristic considerably different from the Butterworth filter. This filter is known as the Chebyshev filter. The form of the Chebyshev filter transfer function is shown below:

124

$$H(j\omega) = \frac{1}{\sqrt{1 + E^2 (C_n)^2}}$$

Equation 5-19

Note that it looks similar to a Butterworth filter transfer function with the exception that there is:

1) A factor in the denominator called E (ripple coefficient), and

2) A term in the denominator called the Chebyshev polynomial. The major difference between the characteristic for a Butterworth low-pass filter and a Chebyshev low-pass filter is that the Butterworth filter has a smooth type of characteristic while the Chebyshev filter has a ripple pattern in its characteristic. Figure 5-6 compares a Butterworth and a Chebyshev filter characteristic. Note that the Butterworth filter characteristic starts at a value of unity at zero frequency and rolls off to a value of zero at infinite frequency. The

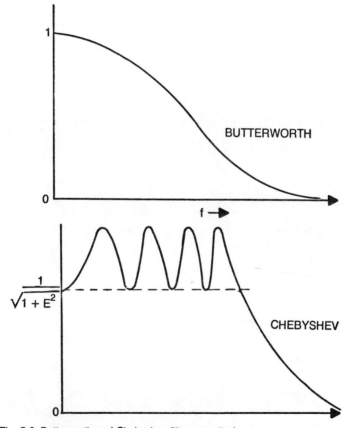

Fig. 5-6. Butterworth and Chebyshev filter magnitude responses.

Chebyshev filter has a response with ripple in it. The ripple increases in frequency as the frequency applied to the filter increases. Eventually the ripple disappears and the filter response decreases to zero.

CHEBYSHEV POLYNOMINAL

Just like a Butterworth filter, the Chebyshev filter has an order, which is called the Chebyshev Polynominal. Table 5-1 gives such polynomials. For example, a zero order Chebyshev filter is really a form of all-pass filter, since the response would be:

$$| H(j\omega) | = \frac{1}{\sqrt{1 + E^2 (1)^2}}$$

$$= \frac{1}{\sqrt{1 + E^2}}$$

Therefore, regardless of frequency the response would always be the same. A first-order Chebyshev filter would have a form of

$$| H (j\omega)| = \frac{1}{\sqrt{1 + E^2 \omega^2}}$$

This would have no ripple in the characteristic.

A second order Chebyshev filter would have the form of

$$| H (j\omega)| = \frac{1}{\sqrt{1 + E^2 (2\omega^2 - 1)^2}}$$

$$= \frac{1}{\sqrt{1 + E^2 (4 \omega^4 - 4 \omega^2 + 1)}}$$

At a frequency of $\omega = 0$, the transfer function would be:

Table 5-1. Chebyshev Polynomials.

Filter Order	Chebyshev Polynominal
0	1
1	ω
2	$2 \omega^2 - 1$
3	$4 \omega^3 - 3 \omega$
4	$8 \omega^4 - 8 \omega^2 + 1$
5	$16 \omega^5 - 20 \omega^3 + 5 \omega$
6	$32 \omega^6 - 48 \omega^4 + 18 \omega^2 - 1$

$$H(j\omega) = \frac{1}{\sqrt{1 + E^2}}$$

The roots of the Polynomial can be determined by setting it equal to zero and solving for frequency as shown

$$2\omega^2 - 1 = 0$$
$$2\omega^2 = 1$$
$$\omega^2 = \tfrac{1}{2}$$
$$\omega = \pm\frac{1}{\sqrt{2}}$$
$$\omega = \pm .707 \text{ radians/second}$$

Since negative frequencies are not valid, we could say that the frequency at which the polynomial will vanish is $\omega = .707$ radians per second. Therefore, we see in Fig. 5-7 that the response of the second-order Chebyshev low-pass filter starts at a value of

$$\frac{1}{\sqrt{1 + E^2}}$$

Then, at a frequency of 0.707 radians per second the response is

$$\frac{1}{\sqrt{1}} = 1$$

After this frequency, the response starts to fall, finally reaching zero at a frequency of infinity. Higher order filters will go through several ripples before the response declines to zero.

Example 5-3. Write the transfer function for a Chebyshev filter if the ripple coefficient is 0.7 and the filter order is 3. From the table, we see that the polynomial is $4\omega^3 - 3\omega$.

The transfer function then becomes

$$H(j\omega) = \frac{1}{\sqrt{1 + E^2 C_n^2}}$$

$$= \frac{1}{\sqrt{1 + (.7)^2 (4\omega^3 - 3\omega)^3}}$$

$$H(j\omega) = \frac{1}{\sqrt{1 + (.49)[16\omega^6 - 24\omega^4 + 9\omega^2]}}$$

$$H(j\omega) = \frac{1}{\sqrt{1 + 7.84\omega^6 - 11.76\omega^4 + 4.41\omega^2}}$$

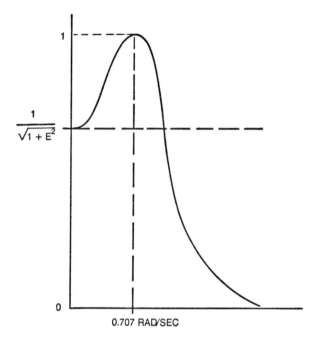

Fig. 5-7. Response of the second-order Chebyshev filter.

$$H(j\omega) = \frac{1}{\sqrt{7.84\omega^6 - 11.76\omega^4 + 4.41\,\omega^2 + 1}}$$

At a frequency of $\omega = 0$, note that the transfer function becomes unity. At a frequency of $\omega = 1$ radian per second, the function becomes unity. In between zero and unity, the function goes through several ripples. Above one radian per second the ripple stops and the function declines to zero. Note, also, that increasing the ripple coefficient, E, increases the size of the ripple.

It is beyond the scope of this book to give a detailed synthesis of the Chebyshev filter or Butterworth filter. Many excellent texts are available that show in great detail how to construct such filters.

CHAPTER SUMMARY

In this chapter, we have learned the basic differences between the construction of low-pass Butterworth and low-pass Chebyshev filters. Basically, the difference in the magnitude response is that the low-pass Butterworth filter has a smooth roll off while the Chebyshev filter is characterized by a ripple response in the pass band.

Problems

1. A third order Butterworth filter has a cutoff frequency of 100 Hz. What is the filter response magnitude at 10,000 Hz?

2. A transfer function has a magnitude of

$$H(j\omega) = \frac{4200}{\sqrt{1+\omega^2}}$$

Extract a stable transfer function from the magnitude equation.

3. Derive the transfer function for a fifth order Butterworth filter.

4. Plot the magnitude and phase response for a fourth order Butterworth filter.

5. A Chebyshev filter has a ripple coefficient of 0.6 and the filter is of the fourth order. Write the magnitude equation for the filter.

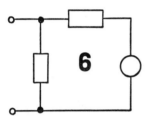

Synthesis of Impedance and Admittance Functions

Being able to look at an equation and estimate what type of circuit it represents is a giant advantage for the student.

This chapter will examine more complicated network equations, and their relationship to the circuits they represent.

WRITING IMPEDANCE EQUATIONS

First of all, let us examine a few circuits and determine their respective impedance equations. We will use S to represent $j\omega$.

Example 6-1. Determine the impedance equations for the circuit in Fig. 6-1.

Solution:

$$Z_{(S)} = R_1 + \frac{R_2 (SL)}{R_2 + SL}$$

$$= 2 + \frac{3 (4S)}{3 + 4S}$$

Writing the equation with a common denominator

$$Z_{(s)} = \frac{2 (3+4S) + 3(4S)}{3 + 4S}$$

$$= \frac{6 + 8S + 12S}{3 + 4S}$$

$$= \frac{6 + 20S}{3 + 4S}$$

To isolate the S in both numerator and denominator terms we obtain

$$Z_{(s)} = \frac{20\left[\dfrac{6}{20} + S\right]}{4\left[\dfrac{3}{4} + S\right]}$$

$$Z_{(s)} = s\left(\frac{0.3 + S}{0.75 + S}\right)$$

Writing the S terms first in the expression, we have

$$Z_{(s)} = s\left(\frac{S + 0.3}{S + 0.75}\right)$$

Example 6-2. Find the impedance equation for the circuit shown in Fig. 6-2.

$$Z_{(s)} = \frac{\left(\dfrac{1}{C_1 S}\right)\left(LS + \dfrac{1}{C_2 S}\right)}{\dfrac{1}{C_1 S} + LS + \dfrac{1}{C_2 S}}$$

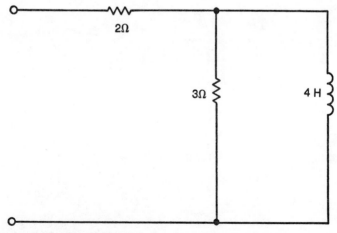

Fig. 6-1. Circuit for example 6-1.

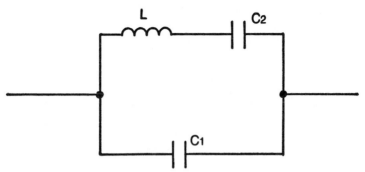

Fig. 6-2. Circuit for example 6-2.

Since $C_1 = C_2$,

$$Z_{(s)} = \frac{\frac{1}{CS}\left(LS + \frac{1}{CS}\right)}{\frac{1}{CS} + LS + \frac{1}{CS}}$$

$$= \frac{\frac{1}{CS}\left(LS + \frac{1}{CS}\right)}{\frac{2}{CS} + LS}$$

$$= \frac{LS + \frac{1}{CS}}{2 + (LS)(CS)}$$

$$= \frac{LS + \frac{1}{CS}}{2 + LCS^2}$$

$$Z_{(s)} = \frac{\frac{LCS^2 + 1}{CS}}{2 + LCS^2}$$

Multiply numerator and denominator by CS:

$$Z_{(s)} = \frac{LCS^2 + 1}{CS(2 + LCS^2)}$$

Isolate the S^2 in the numerator and denominator by dividing by LC:

$$Z_{(s)} = \frac{\dfrac{LCS^2 + 1}{LC}}{CS\left(\dfrac{2}{LC} + S^2\right)}$$

$$= \frac{S^2 + \dfrac{1}{LC}}{CS\left(S^2 + \dfrac{2}{LC}\right)}$$

$$= \frac{S^2 + \dfrac{1}{0.04}}{0.1S\left(S^2 + \dfrac{2}{0.04}\right)}$$

$$= \frac{S^2 + 25}{0.1S\,(S^2 + 50)}$$

$$= \frac{10\,(S^2 + 25)}{S\,(S^2 + 50)}$$

Example 6-3. Find the admittance for the network shown in Fig. 6-3.

Solution:

First, write the impedance equation for the circuit:

$$Z_{(s)} = SL + R_1 + \frac{1}{SC + G_2}$$

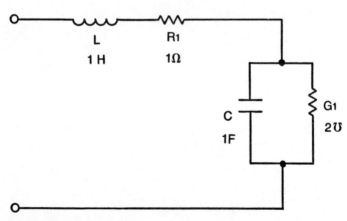

L
1 H

R1
1Ω

C
1F

G1
2℧

Fig. 6-3. Circuit for example 6-3.

$$= S(1) + (1) + \frac{1}{S(1) + 2}$$

$$= S + 1 + \frac{1}{S + 2}$$

$$= \frac{(S+1)(S+2) + 1}{S + 2}$$

$$= \frac{S^2 + S + 2S + 2 + 1}{S + 2}$$

$$= \frac{S^2 + 3S + 3}{S + 2}$$

Since impedance and admittance are reciprocal quantities,

$$Y_{(s)} = \frac{1}{Z_{(s)}} = \frac{1}{\dfrac{S^2 + 3S + 3}{S + 2}}$$

$$Y_{(s)} = \frac{S + 2}{S^2 + 3S + 3}$$

Example 6-4. What is the impedance equation for the circuit shown in Fig. 6-4?

$$Z_{(s)} = \frac{(SL + R)\left(\dfrac{1}{CS}\right)}{SL + R + \dfrac{1}{CS}}$$

$$= \frac{(S(10) + 3)\left(\dfrac{1}{4S}\right)}{S(10) + 3 + \dfrac{1}{4S}}$$

$$= \frac{10S + 3}{4S\left(10S + 3 + \dfrac{1}{4S}\right)}$$

$$= \frac{10S + 3}{40S^2 + 12S + 1}$$

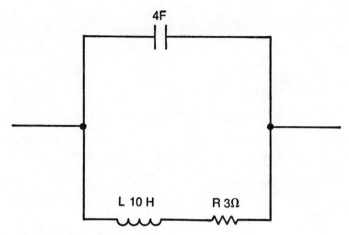

Fig. 6-4. Circuit for example 6-4.

PARTIAL FRACTIONS

One of the major tools of network synthesis is called partial fractions. A simple example will serve to illustrate.

Example 6-5. In example 6-1, we wrote the impedance equation for an RL circuit. Now let us see if we can obtain the original circuit from the equation.

The equation was

$$Z_{(s)} = \frac{5 (S + 0.3)}{S + 0.75}$$

We can write this equation in two parts by using the following technique:

A fraction of the form

$$Z_{(s)} = \frac{S + a}{S (S + b)}$$

can be written as

$$Z_{(s)} = \frac{K_1}{S} + \frac{K_2}{S + b}$$

However, we note that there is no S term other than S + 0.75 in the denominator. We can supply an S term as follows:

$$Z_{(s)} = \frac{5 (S + 0.3)}{S (S + 0.75)}$$

$$= \frac{K_1}{S} + \frac{K_2}{S + 0.75}$$

$$\frac{5(S + 0.3)}{S(S + 0.75)} = \frac{K_1}{S} + \frac{K_2}{S + 0.75}$$

Multiplying both sides of the equation by S (S + 0.75) we obtain,

$$5(S + 0.3) = K_1(S + 0.75) + K_2(S)$$

To determine K_2, let S equal -0.75. This will cause (S + 0.75) to vanish, and therefore K_1 is multiplied by zero. We then get the following:

$$5(-.75 + 0.3) = K_2(-.75)$$
$$5(-0.45) = K_2(-.75)$$
$$K_2 = \frac{5(0.45)}{0.75} = 3$$

If S = 0, we then obtain the following equation:

$$5(0.3) = K_1(0 + 0.75) + K_2(0)$$
$$1.5 = .75 K_1$$
$$K_1 = \frac{1.5}{0.75}$$
$$= 2$$

We may then write,

$$\frac{Z_{(s)}}{S} = \frac{2}{S} + \frac{3}{S + 0.75}$$

Since we supplied an S in the denominator of the impedance function, we now multiply both sides of the equation by S to obtain the following relation:

$$Z_{(s)} = 2 + \frac{3S}{S + 0.75}$$

Since the equation is an impedance, and we have a quantity of 2 added to $\frac{3S}{S + 0.75}$ we know that both of these must be impedances. The 2 means 2 ohms, while the quantity $\frac{3S}{S + 0.75}$ must represent ohms also.

We know that an inductor and resistor in parallel have an impedance equation of:

$$Z = \frac{RLS}{SL + R}$$

This can be written as

$$Z = \frac{RS}{S + \dfrac{R}{L}}$$

By comparison with $\dfrac{3S}{S + 0.75}$, the R must be equal to 3 ohms, and the $\dfrac{R}{L}$ must be equal to 0.75. Therefore,

$$\frac{R}{L} = 0.75$$

$$\frac{3}{L} = 0.75$$

$$L = \frac{3}{0.75}$$

$$L = 4H$$

The answer to this problem is a circuit made up of 2-ohm resistor in series with a parallel circuit composed of a 3-ohm resistor and a 4-henry inductor.

The resultant circuit is shown in Fig. 6-5.

Example 6-6. Suppose we had been given the equation determined in example 6-2. We would like to derive the circuit.

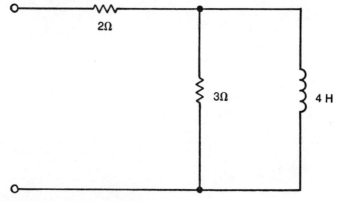

Fig. 6-5. Circuit for example 6-5.

Solution:

The equation for $Z_{(s)}$ is

$$Z_{(s)} = \frac{10 (S^2 + 25)}{S (S^2 + 50)}$$

Then write the following relation:

$$Z_{(s)} = \frac{10 (S^2+25)}{S (S^2+50)} = \frac{A}{S} + \frac{BS}{S^2 + 50}$$

Multiply both sides of the equation by $S(S^2 + 50)$ and obtain

$$10 (S^2 + 25) = A (S^2 + 50) + BS^2$$

If we allow S^2 to become zero, we can write

$$10 (0 + 25) = A (0 + 50) + B (0)$$
$$250 = 50 A$$
$$A = \frac{250}{50}$$
$$A = 5$$

If S^2 is $- 50$, we obtain

$$10 (-50 + 25) = A (-50 +50) + B (-50)$$
$$10 (-25) = -50 B$$
$$-250 = -50 B$$
$$B = \frac{250}{50}$$
$$B = 5$$

We can then write the equation for $Z_{(s)}$:

$$Z_{(s)} = \frac{5}{S} + \frac{5S}{S^2 + 50}$$

The quantity $\frac{5}{S}$ represents a capacitive reactance of $\frac{1}{CS}$, where C would represent 0.2 farads.

The quantity $\frac{5S}{S^2 + 50}$ represents an LC parallel circuit. The circuit would have an impedance given by

$$\frac{(LS) \left(\frac{1}{CS}\right)}{LS + \frac{1}{CS}}$$

which is equal to

$$\frac{LS}{LCS^2 + 1}$$

This expression can be represented as

$$\frac{\dfrac{LS}{LC}}{\dfrac{LCS^2}{LC} + \dfrac{1}{LC}} = \frac{\left(\dfrac{S}{C}\right)}{S^2 + \dfrac{1}{LC}}$$

By comparison with the equation

$$\frac{5\,S}{S^2 + 50}$$

we see that 5 would represent $\dfrac{1}{C}$ and the 50 represent $\dfrac{1}{LC}$.

Solving for C, we would have

$$\frac{1}{C} = 5$$

$$C = \frac{1}{5}$$

$$C = 0.2 \text{ farads.}$$

Solving for L, we would obtain

$$\frac{1}{LC} = 50$$

$$L = \frac{1}{50\,C}$$

$$= \frac{1}{50\,(.2)}$$

$$= \frac{1}{10}$$

$$= 0.1 \text{ henry}$$

The resulting circuit is that shown in Fig. 6-6, which is equivalent to that in Fig. 6-2.

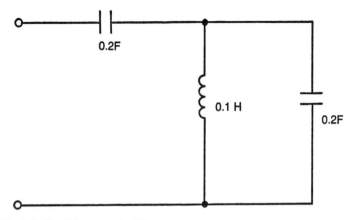

Fig. 6-6. Circuit for example 6-6.

FOSTER'S SYNTHESIS

This is a method of evaluating a function so that it represents either of the following:

a) a series circuit composed of an inductor, capacitor, and a group of parallel circuits each composed of a capacitor and inductor this is Foster's first approximation.

b) a parallel circuit composed of the following branches: 1. capacitor, 2. inductor, 3. capacitor and inductor. This is Foster's second approximation.

This is Foster's second approximation.

Foster's first approximation takes an impedance equation such as that given by

$$Z_{(s)} = K_a S + \frac{K_b}{S} + \frac{2 K_c S}{S^2 + \omega_c^2} + \frac{2 K_d S}{S^2 + \omega_d^2} \qquad \text{Equation 6-1}$$

The resulting circuit is shown in Fig. 6-7. As an example, let us determine Foster's first approximation for the following equation:

$$Z_{(s)} = \frac{2 S^4 + 8S^2 + 6}{S^3 + 2S}$$

Solution:

1. Divide the numerator by the denominator.

140

$$S^3 + 2S \overline{\smash{\big)}\ 2S^4 + 8S^2 + 6} \quad \frac{2S}{}$$
$$\underline{2S^4 + 4S^2}$$
$$4S^2 + 6$$

We can then write the impedance as

$$Z_{(s)} = 2S + \frac{4S^2 + 6}{S^3 + 2S}$$

2. If we then apply the technique of partial fractions we have

$$Z_{(s)} = K_a S + \frac{A}{S} + \frac{BS}{S^2 + \omega_c^2}$$

$$\frac{2S^4 + 8S^2 + 6}{S^3 + 2S} = 2S + \frac{A}{S} + \frac{BS}{S^2 + 2}$$

$$2S + \frac{4S^2 + 6}{S^3 + 2S} = 2S + \frac{A}{S} + \frac{BS}{S^2 + 2}$$

By comparison,

$$\frac{4S^2 + 6}{S^3 + 2S} = \frac{A}{S} + \frac{BS}{S^2 + 2}$$

If we multiply each side of the equation by $S^3 + 2S$, we obtain

$$4S^2 + 6 = A (S^2 + 2) + BS^2$$

If $S^2 = -2$, we obtain

$4 (-2) + 6$	$= A (-2 + 2) + B (-2)$
-2	$= -2B$
B	$= 1$

Setting $S^2 = 0$, we have

$4 (0) + 6$	$= A (0 + 2) + B (0)$
6	$= 2 A$
A	$= \frac{6}{2}$
A	$= 3$

3. We obtain the following impedance equation using the solution from the partial fractions.

$$Z_{(s)} = 2S + \frac{3}{S} + \frac{S}{S^2 + 2}$$

Comparing with equations, we have

$$Z_{(s)} = K_a S + \frac{K_b}{S} + \frac{2K_c S}{S^2 + \omega_c^2}$$

where: $K_a = 2$

$K_b = 3$

$2K_c = 1$

$\omega_c^2 = 2$

Using Fig. 6-7, we have

$$L_1 = K_a = 2H$$

$$C_1 = \frac{1}{K_b} = \frac{1}{3} F$$

$$C_2 = \frac{1}{2K_c}$$

$$= \frac{1}{1}$$

$$= 1F$$

$$L_2 = \frac{2K_d}{\omega_c^2}$$

$$= \frac{1}{2} H$$

This circuit is shown in Fig. 6-8.

Foster's second approximation takes an impedance, $Z_{(s)}$, and turns it into the equivalent parallel network. The steps are as follows:

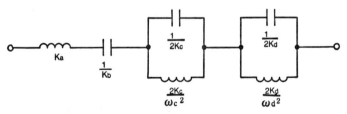

Fig. 6-7. Foster's first approximation circuit.

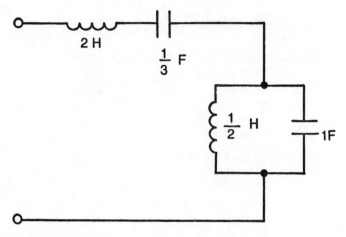

Fig. 6-8. Resultant circuit for finding Foster's first approximation for the equation:
$Z(s) = \dfrac{2s^4 + 8s^2 + 6}{s^3 + 2s}$

1. Take the reciprocal of $Z_{(s)}$, obtaining the admittance:

$$Y_{(s)} = \frac{1}{Z_{(s)}}$$

2. Finding $Y_{(s)}$, we obtain an expression

$$Y_{(s)} = K'_a S + \frac{K'_b}{S} + \frac{2K'_c S}{S^2 + \omega_c^2} + \frac{2K'_d S}{S^2 + \omega_d^2} \qquad \text{Equation 6-2}$$

The circuit which such an equation represents is shown in Fig. 6-9.
Example 6-7. Find Foster's second approximation for the following equation:

$$Z_{(s)} = \frac{2S^4 + 8S^2 + 6}{S^3 + 2S}$$

Write the reciprocal of $Z_{(s)}$

$$Y_{(s)} = \frac{1}{Z_{(s)}}$$

$$= \frac{1}{\dfrac{2S^4 + 8S^2 + 6}{S^3 + 2S}}$$

$$= \frac{S^3 + 2S}{2S^4 + 8S^2 + 6}$$

$$= \frac{1}{2}(S) \frac{S^2 + 2}{(S^2 + 3)(S^2 + 1)}$$

Apply your knowledge of partial fractions and obtain:

$$Y_{(s)} = \frac{1}{2} \frac{S(S^2 + 2)}{(S^2 + 3)(S^2 + 1)} = \frac{1}{2} \left(\frac{AS}{S^2+3} + \frac{BS}{S^2+1} \right)$$

Multiply through by $2(S^2 + 3)(S^2 + 1)$:

$$S(S^2 + 2) = AS(S^2 + 1) + BS(S^2 + 3)$$

Suppose $S^2 = -1$, we obtain

$$S(-1 + 2) = AS(-1 + 1) + BS(-1 + 3)$$

$$S = BS(2)$$

$$1 = 2B$$

$$B = \frac{1}{2}$$

If $S^2 = -3$, we obtain

$$S(-3 + 2) = AS(-3 + 1) + BS(-3 + 3)$$

$$-S = -2AS$$

$$1 = 2A$$

$$A = \frac{1}{2}$$

3. Place the values of A and B into the impedance equation and obtain

$$Y_{(s)} = \frac{1}{2} \left(\frac{\frac{1}{2}S}{S^2 + 3} + \frac{\frac{1}{2}S}{S^2 + 1} \right)$$

We can rewrite the equation like this:

$$Y_{(s)} = \frac{\frac{1}{4}S}{S^2 + 3} + \frac{\frac{1}{4}S}{S^2 + 1}$$

By referring to Fig. 6-9 and equation 6-2, we can write the following:

$$2K'_c = \frac{1}{4}$$

$$K'_c = \frac{1}{8}$$

$$\omega_c^2 = 3$$

$$2K'_d = \frac{1}{4}$$

$$K_d' = \frac{1}{8}$$

$$\omega_d^2 = 1$$

Using this information, we can write,

$$L_1 = \frac{1}{2 K_c'} = \frac{1}{\frac{1}{4}} = 4 \text{ henries}$$

$$C_1 = \frac{2 K_c'}{\omega_c^2} = \frac{\frac{1}{4}}{(3)} = \frac{1}{12} \text{ farad}$$

$$L_2 = \frac{1}{2 K_d'} = \frac{1}{\frac{1}{4}} = 4 \text{ henries}$$

$$C_2 = \frac{2 K_d'}{\omega_d^2} = \frac{\frac{1}{4}}{(1)} = \frac{1}{4} \text{ farad}$$

The circuit for this equation is shown in Fig. 6-10.

CAUER'S SYNTHESIS

Like Foster's synthesis, this technique also has two forms. These are named Cauer I and Cauer II.

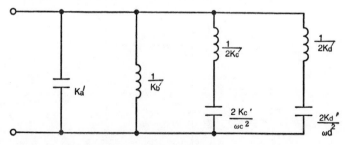

Fig. 6-9. Foster's second approximation circuit.

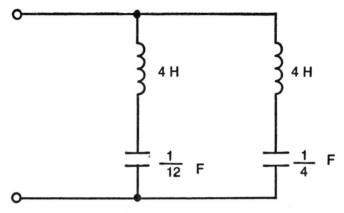

Fig. 6-10. Solution to example 6-7.

Cauer I

In this technique, the impedance equation which is synthesized can be written in the form

$$Z_{(s)} = a_1 S + \cfrac{1}{a_2 S + \cfrac{1}{a_3 S + \cfrac{1}{a_4 S}}} \qquad \text{Equation 6-3}$$

The circuit such an equation represents is given in Fig. 6-11. As an example:

$$Z_{(s)} = \frac{2 S^4 + 8S^2 + 6}{S^3 + 2S}$$

The steps in determining the equivalent circuit are as follows:

1. Divide numerator by denominator.

$$S^3 + 2S \overline{)\begin{array}{l} 2S^4 + 8S^2 + 6 \\ \underline{2S^4 + 4S^2} \\ 4S^2 + 6 \end{array}}$$

2. Divide original divisor by remainder

$$4S^2 + 6 \overline{)\begin{array}{l} \dfrac{S}{4} \\ S^3 + 2S \\ \underline{S^3 + 1.5S} \\ 0.5S \end{array}}$$

3. Divide divisor of step two by the remainder of last step.

146

$$0.55 \overline{\smash{\big)}\begin{array}{l} 8S \\ 4S^2 + 6 \\ \underline{4S^2} \\ 6 \end{array}}$$

4. Divide divisor of last step by remainder of last step.

$$6 \overline{\smash{\big)}\begin{array}{l} S/12 \\ 0.5S \\ \underline{0.5S} \end{array}}$$

5. There is no remainder in step four and the division is stopped at this point.

6. Refer to Equation 6-3 and note the following:

a. For step one, the quotient 2S represents $(^{\alpha}1)S$.

b. For step two, the quotient $\dfrac{S}{4}$ represents $(^{\alpha}2)S$.

c. For step three, the quotient 8S represents $(^{\alpha}3)S$.

d. For step four, the quotient $\dfrac{S}{12}$ represents $(^{\alpha}4)S$.

Referring to Fig. 6-11, we can then say,

1. $L_1 = {}^{\alpha}1 = 2$ henries

2. $C_1 = {}^{\alpha}2 = \dfrac{1}{4}$ farad

3. $L_2 = {}^{\alpha}3 = 8$ henries

4. $C_2 = {}^{\alpha}4 = 1/12$ farad

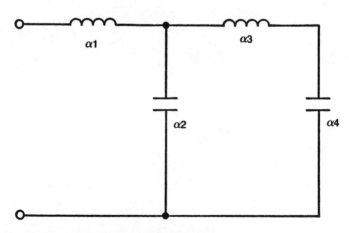

Fig. 6-11. Cauer's first approximation circuit.

Figure 6-12 shows the equivalent circuit which represents the equation

$$Z_{(s)} = \frac{2S^4 + 8S^2 + 6}{S^3 + 2S}$$

Cauer II

In this technique, the impedance equation which is synthesized can be written in the form

$$Z_{(s)} = \frac{1}{b_1 S} + \cfrac{1}{\cfrac{1}{b_2 S} + \cfrac{1}{\cfrac{1}{b_3 S} + \cfrac{1}{b_4 S}}}$$

The circuit is shown in Fig. 6-13.

As an example, let us draw a Cauer II representation from the following:

$$Z_{(s)} = \frac{2S^4 + 8S^2 + 6}{S^3 + 2S}$$

First, arrange the numerator and denominator in ascending powers of S:

$$Z_{(s)} = \frac{6 + 8S^2 + 2S^4}{2S + S^3}$$

Then divide the numerator by the denominator:

$$
\begin{array}{r}
\dfrac{3}{S} \\[4pt]
2S + S^3 \overline{\big)\, 6 + 8S^2 + 2S^4} \\
\underline{6 + 3S^2} \\
5S^2 + 2S^4
\end{array}
$$

Divide the original divisor by the remainder:

$$
\begin{array}{r}
\dfrac{2}{5S} \\[4pt]
5S^2 + 2S^4 \overline{\big)\, 2S + S^3} \\
2S + \underline{\dfrac{4}{5}\, S^3} \\
\hline
\dfrac{S^3}{5}
\end{array}
$$

Divide the divisor of step two by the remainder of the last step:

$$\cfrac{S^3}{5} \Big) \overline{\begin{array}{c} \cfrac{25}{S} \\ 5S^2 + 2S^4 \\ \underline{5S^2} \\ 2S^4 \end{array}}$$

Divide the divisor of the previous step by the remainder of that step:

$$2S^4 \Big) \overline{\begin{array}{c} \cfrac{1}{10S} \\ \cfrac{S^3}{5} \\ \underline{\cfrac{S^3}{5}} \end{array}}$$

There is no remainder in this last ste, so we step division. Note the following as we refer to Equation 6-4.

    a. For step one, the quotient $\dfrac{3}{5}$ represents $\dfrac{1}{b_1 S}$.

    b. For step two, the quotient $\dfrac{2}{5S}$ represents $\dfrac{1}{b_2 S}$.

    c. For step three, the quotient $\dfrac{25}{S}$ represents $\dfrac{1}{b_3 S}$.

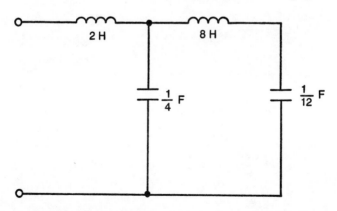

Fig. 6-12. Resultant circuit for finding Cauer's first approximation for the equation:
$$Z(s) = \frac{2 s^4 + 8 s^2 + 6}{s^3 + 2s}$$

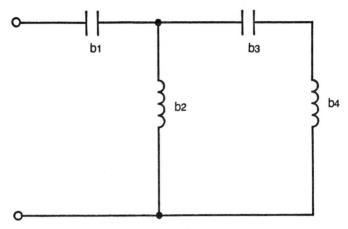

Fig. 6-13. Cauer's second approximation circuit.

    d. For step four, the quotient $\dfrac{1}{10S}$ represents $\dfrac{1}{b_4 S}$.

Referring to Fig. 6-13 we can then say,

1. $\dfrac{3}{S} = \dfrac{1}{b_1 S}$

    $b_1 = 1/3$

    $C_1 = b_1 = 1/3$ farad

2. $\dfrac{2}{5S} = \dfrac{1}{b_2 S}$

    $b_2 = \dfrac{5}{2}$

    $L_1 = b_2 = \dfrac{5}{2}$ henries

3. $\dfrac{25}{S} = \dfrac{1}{b_3 S}$

    $b_3 = \dfrac{1}{25}$

    $C_2 = b_3 = \dfrac{1}{25}$ farad

4. $\dfrac{1}{10S} = \dfrac{1}{b_4 S}$

$b_4 = 10$

$L_2 = b_4 = 10$ henries

Figure 6-14 shows the equivalent circuit, which represents the equation

$$Z_{(s)} = \dfrac{2S^4 + 8S^2 + 6}{S^3 + 2S^2}$$

Example 6-8. Synthesize the function:

$$Z_{(s)} = \dfrac{S^5 + 3S^3 + 2S}{2S^6 + 11S^4 + 16S^2 + 4}$$

$$Y_{(s)} = \dfrac{1}{Z_{(s)}} = \dfrac{2S^6 + 11S^4 + 16S^2 + 4}{S^5 + 3S^3 + 2S}$$

We now make a division

$$
\begin{array}{r}
2S \\
S^5 + 3S^3 + 2S \overline{\big)\; 2S^6 + 11S^4 + 16S^2 + 4} \\
2S^6 + 6S^4 + 4S^2 \\
\hline
5S^4 + 12S^2 + 4
\end{array}
$$

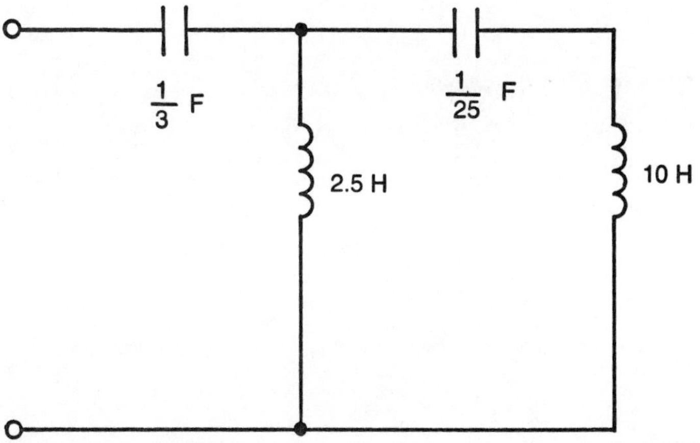

Fig. 6-14. Resultant circuit for finding Cauer's second approximation for the equation:

$$Z(s) = \dfrac{2s^4 + 8s^2 + 6}{S^3 + 2s}$$

151

$$Y_{(s)} = 2S + \frac{5S^4 + 12S^2 + 4}{S^5 + 3S^3 + 2S}$$

$$= 2S + \frac{5S^4 + 12S^2 + 4}{S(S^4 + 3S^2 + 2)}$$

$$= 2S + \frac{5S^4 + 12S^2 + 4}{S(S^2 + 1)(S^2 + 2)}$$

Now see if it is possible to factor the numerator of the fraction. We can write

$$\frac{5S^4 + 12S^2 + 4}{S(S^2 + 1)(S^2 + 2)} = \frac{(5S^2 + 2)(S^2 + 2)}{S(S^2 + 1)(S^2 + 2)}$$

$$= \frac{5S^2 + 2}{S(S^2 + 1)}$$

Using partial fractions, we can then write

$$\frac{5S^2 + 2}{S(S^2 + 1)} = \frac{A}{S} + \frac{BS}{S^2 + 1}$$

Multiplying through by $S(S^2 + 1)$ we obtain,

$$5S^2 + 2 = A(S^2 + 1) + BS^2$$

If we let $S^2 = -1$, we obtain

$$5(-1) + 2 = A(-1 + 1) + B(-1)$$
$$-3 = -1B$$
$$B = 3$$

If we let $S^2 = 0$, we obtain

$$5(0) + 2 = A(0 + 1) + B(0)$$
$$2 = A$$

We then write

$$Y_{(s)} = 2S + \frac{2}{S} + \frac{3S}{S^2 + 1}$$

Refer to Fig. 6-9. The components are found as follows:

$$Ka' = 2, \quad C_1 = 2 \text{ farads}$$

$$Kb' = 2, \quad \frac{1}{Ko'} = L_1 = 1/2 \text{ henry}$$

$$2Kc' = 3, \frac{1}{2K'c} = L_2 = 1/3 \text{ henry}$$

$$(\omega c')^2 = 1, \frac{2Kc'}{\omega c'^2} = 3 \text{ farads}$$

The resulting circuit is shown in Fig. 6-15.

Example 6-9. Synthesize the function

$$Z(s) = \frac{(S + 1)(S + 4)}{(S + 3)(S + 5)}$$

First of all, supply an S in the denominator. The partial fraction result is

$$\frac{Z(s)}{S} = \frac{A}{S} + \frac{B}{S + 3} + \frac{C}{S + 5}$$

Then multiply through by $(S + 3)(S + 5)$ S

$$(S+1)(S+4) = A(S+3)(S+5) + BS$$
$$(S+5) + CS(S+3)$$

If we let $S = 0$, we obtain,

$$1(4) = A(3)(5) + B(0)(5) + C(0)(3)$$
$$4 = 15A$$
$$A = \frac{4}{15}$$

If we let $S = -5$, we obtain

$$(-5+1)(-5+4) = A(-5+3)(-5+5) + B(-5)(-5+5) +$$
$$C(-5)(-5+3)$$
$$(-4)(-1) = 0 \quad + 0 \quad + 10C$$
$$C \qquad = \frac{4}{10} = \frac{2}{5}$$

If we let $S = -3$, we obtain

$$(-3+1)(-3+4) = A(-3+3)(-3+5) + B(-3)(-3+5) +$$
$$C(-3)(-3+3)$$
$$(-2)(1) = -6B$$

$$B \qquad = \frac{2}{6}$$

$$B \qquad = \frac{1}{3}$$

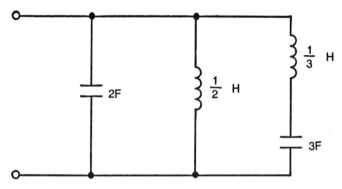

Fig. 6-15. Solution to example 6-8.

Then put the values of A, B, and C into the expression for $Z_{(s)}$. We have

$$\frac{Z_{(s)}}{S} = \frac{\frac{4}{15}}{S} + \frac{\frac{1}{3}}{S+3} + \frac{\frac{2}{5}}{S+5}$$

Multiplying both sides of the equation by S, we obtain

$$Z_{(s)} = \frac{4}{15} + \frac{\frac{1}{3}S}{S+3} \qquad \frac{\frac{2}{5}S}{S+5}$$

By examination, we see that $Z_{(s)}$ is the sum of three impedances. The first one being a resistor of value $\frac{4}{15}$ ohm.

However, the second and third expressions represent complex impedances. The quantity $\frac{\frac{1}{3}S}{S+3}$ represents an inductor of $\frac{1}{9}$ henry and a resistor of $\frac{1}{3}$ ohm in parallel. The quantity $\frac{\frac{2}{5}S}{S+5}$ represents a resistor of $\frac{2}{5}$ ohm and an inductor of $\frac{2}{25}$ henry in parallel. We can see the resulting circuit in Fig. 6-16.

Now, suppose that the function had been an admittance, namely,

154

$$Y_{(s)} = \frac{(S+1)(S+4)}{(S+3)(S+5)}$$

The procedure would be the same in regards to partial fractions, giving

$$Y_{(s)} = \frac{4}{15} + \frac{\frac{1}{3}S}{S+3} + \frac{\frac{2}{5}S}{S+5}$$

By examination, we see that $Y_{(s)}$ is the sum of three admittances. The first is a conductance of $\frac{4}{15}$ mho. The second expression represents a capacitor and resistor in series. To determine the values, we think of $\frac{\frac{1}{3}S}{S+3}$ as the reciprocal of an impedance, Namely,

$$\frac{\frac{1}{3}S}{S+3} = \frac{\frac{1}{3}(S)}{\frac{1}{3}(S)} \cdot \frac{1}{\frac{S}{\frac{1}{3}(S)} + \frac{3}{\frac{1}{3}(S)}}$$

$$= \frac{1}{\frac{1}{\frac{1}{3}} \quad \frac{3}{\frac{1}{3}S}}$$

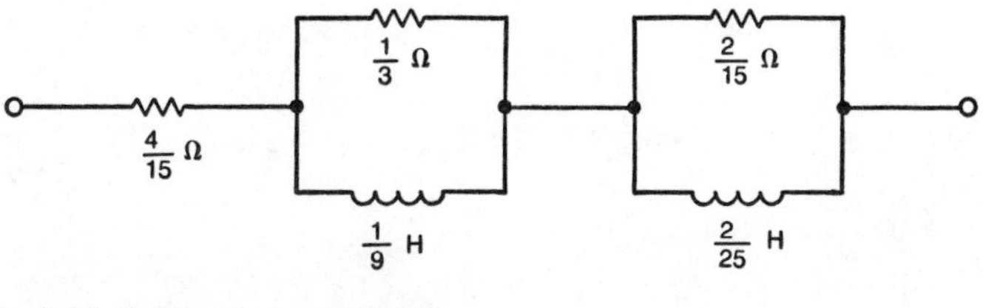

Fig. 6-16. Solution to example 6-9.

$$= \cfrac{1}{3 + \cfrac{9}{S}}$$

$$= \cfrac{1}{3 + \cfrac{1}{\cfrac{1}{9}S}}$$

This represents a capacitor of $\frac{1}{9}$ farad in series with a 3-ohm resistor.

The quantity $\cfrac{\frac{2}{5}(S)}{S + 5}$ also represents a resistor in series with a capacitor. If we think of the expression as a reciprocal of an impedance we have

$$\cfrac{\left(\frac{2}{5}\right)S}{S + 5} = \cfrac{\left(\frac{2}{5}\right)S}{\cfrac{(2/5)S}{\cfrac{S}{\left(\frac{2}{5}\right)} + \cfrac{5}{\left(\frac{2}{5}\right)}S}}$$

$$= \cfrac{1}{\cfrac{5}{2} + \cfrac{25}{2S}}$$

$$= \cfrac{1}{\cfrac{5}{2} + \cfrac{1}{\left(\frac{2}{25}\right)S}}$$

This represents a resistor of 2.5 ohms and a capacitor of $\frac{2}{25}$ farads in series. The complete circuit is shown in Fig. 6-7.

Problems

1. Find Cauer's first and second form for the following expressions:

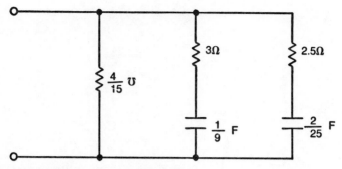

Fig. 6-17. Second solution to example 6-9.

a. $Z(s) = \dfrac{(S^2 + 2)(S^2 + 4)}{S(S^2 + 3)}$

b. $Z(s) = \dfrac{19(S^2 + 8)}{S(S^2 + 10)}$

2. Determine Foster's first and second form for the following expressions:

a. $Z(s) = \dfrac{S(S^4 + 3S^2 + 2)}{S^6 + 12S^4 + 20S^2 + 8}$

b. $Z(s) = \dfrac{12(S^2 + 30)}{S(S^2 + 80)}$

c. $Z(s) = \dfrac{18S^4 + 20S^2 + 21}{S^3 + 3S}$

3. Synthesize:

a. $Y(s) = \dfrac{4(S + 4)(S + 6)}{(S + 2)(S + 8)}$

b. $Z(s) = 16 + \dfrac{9S}{4 + 3S}$

4. Synthesize the following expression as an impedance as well as an admittance.

$$F(s) = \dfrac{8(S + 4)(S + 2)}{(S + 3)(S + 10)}$$

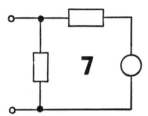

Active Network Synthesis

So far, we have addressed the subject of network synthesis from a passive viewpoint. Now let us see how active devices can be used in conjunction with passive components to obtain solutions to synthesis problems.

In an earlier chapter, we examined various types of transfer functions. Some of these functions could be very easily synthesized if we could simply multiply basic transfer functions directly. For example, suppose we desire to obtain a transfer function of the form

$$H(S) = \frac{2}{S^2 + 3S + 2}$$

It only seems correct that we could obtain such a function by multiplying two functions together like this:

$$\frac{2}{S^2 + 3S + 2} = \frac{1}{(S + 1)} \times \frac{2}{(S + 2)}$$

Each one of the individual transfer functions could represent a basic low-pass filter. The circuits of Figs. 7-1 and 7-2 represent the transfer functions of $\frac{1}{S+1}$ and $\frac{2}{S+2}$ respectively. However, if we place the two circuits in cascade (Fig. 7-3A), we obtain the transfer function by the following method:

$$E_i = I_1\left(R_1 + \frac{1}{SC_1}\right) - I_2\left(\frac{1}{SC_1}\right)$$

$$0 = -I_1 \frac{(1)}{SC_1} + I_2\left(\frac{1}{SC_1} + R_2 + \frac{1}{SC_2}\right)$$

Since $R_1 = 1$ ohm, $C_1 = 1$ farad, $R_2 = \frac{1}{2}$ ohm, and $C_2 = 1$ farad,

$$E_i = I_1\left(1 + \frac{1}{S}\right) - I_2\left(\frac{1}{S}\right)$$

$$0 = -I_1\left(\frac{1}{S}\right) + I_2\left(\frac{1}{S} + \frac{1}{2} + \frac{1}{S}\right)$$

The Δ of the system becomes

$$\Delta = \left(1 + \frac{1}{S}\right)\left(\frac{2}{S} + \frac{1}{2}\right) - \left(\frac{-1}{S}\right)\left(\frac{-1}{S}\right)$$

$$= \frac{2}{S} + \frac{1}{2} + \frac{2}{S^2} + \frac{1}{2S} - \frac{1}{S^2}$$

$$= \frac{1}{S^2} + \frac{4S + S}{2\,S^2} + \frac{1}{2}$$

Finding ΔI_2, we obtain

$$\Delta I_2 = 1 + \frac{1}{S}\ ,\ E_i - \frac{1}{S}\ ,\ 0$$

$$= -(E_i)\left(\frac{-1}{S}\right)$$

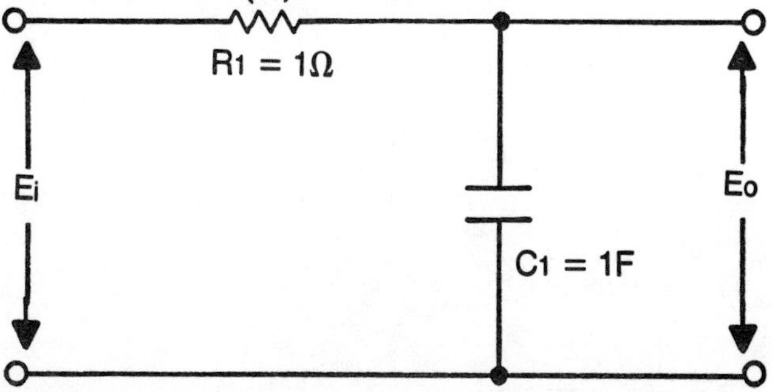

Fig. 7-1. Low-pass circuit having a cutoff of 1 radian / second.

159

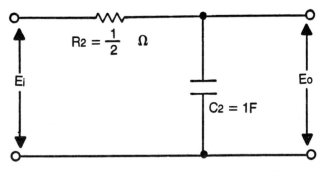

Fig. 7-2. Low pass circuit having a cutoff of 2 radians / second.

$$= \frac{E_i}{S}$$

The output voltage is the product of I_2 and $\frac{(1)}{S}$

$$E_2 = \frac{\Delta I_2}{\Delta} \left(\frac{1}{S}\right)$$

$$E_2 = \frac{\dfrac{E_i}{S}}{\dfrac{1}{S^2} + \dfrac{4S + S}{2S^2} + \dfrac{1}{2}} \left(\frac{1}{S}\right)$$

$$= \frac{E_i \left(\dfrac{1}{S^2}\right)}{\dfrac{1}{S^2} + \dfrac{5S}{2S^2} + \dfrac{1}{2}}$$

$$= \frac{E_i}{1 + \dfrac{5}{2} S + \dfrac{S^2}{2}}$$

$$E_o = \frac{2 E_i}{2 + 5S + S^2}$$

$$\frac{E_o}{E_i} = \frac{2}{S^2 + 5S + 2}$$

This transfer function is not what we desired. The reason is that there is no buffering action between stages. If there were a buffer, we could obtain the transfer function by simply having one low-pass network at the input of a buffer amplifier, and another low-pass network at the output. For example, an integrated-circuit operational amplifier could be wired in the non-inverting mode to act as a buffer. The input impedance would be extremely large, preventing any loading of the input network while, at the same time, the output impedance is usually quite low, and does not effect the circuit connected to its terminal. Figure 7-3B indicates just such a situation where a buffer is being used.

Now, let us examine another problem. Suppose you were given the following transfer function and asked to build the circuit it represents:

$$H(S) = \frac{40}{S^2 + 12S + 20}$$

First of all, see if we can easily factor the equation. In this case it is possible. It factors into

$$H(S) = 40\left(\frac{1}{S + 2}\right)\left(\frac{1}{S + 10}\right)$$

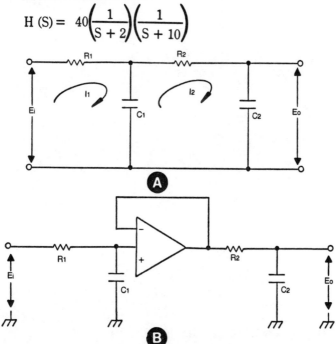

Fig. 7-3. Two low pass filters in cascade (A), and with a buffer (B).

The quantity $\dfrac{1}{S+2}$ is related to a low-pass filter with a cutoff frequency of two radians per second. This can be seen by the form of a low-pass filter than is shown in Fig. 7-4A.

$$H(j\omega) = \frac{\dfrac{1}{j\omega C}}{\dfrac{1}{j\omega C} + R}$$

$$= \frac{1}{1 + j\omega\, CR}$$

$$= \frac{1}{1 + \dfrac{j\omega}{\dfrac{1}{CR}}}$$

$$= \frac{1}{1 + \dfrac{j\omega}{\omega C}}$$

where ωC = cutoff frequency

Or we can write

$$H(j\omega) = \frac{\dfrac{1}{CR}}{\dfrac{1}{CR} + j\omega}$$

If we let $j\omega = S$, then

$$H(S) = \frac{\dfrac{1}{CR}}{\dfrac{1}{CR} + S}$$

By inspection with the form of the transfer function, we see that the cutoff must be two radians per second.

However, to make the transfer function complete, we should have a 2 in its numerator as shown below:

162

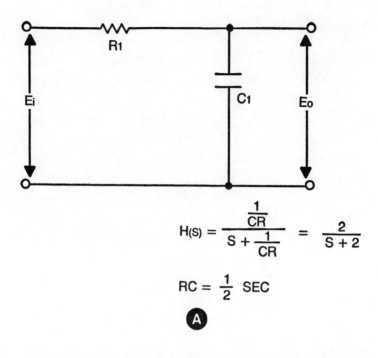

$$H_{(S)} = \frac{\frac{1}{CR}}{S + \frac{1}{CR}} = \frac{2}{S + 2}$$

$$RC = \frac{1}{2} \text{ SEC}$$

A

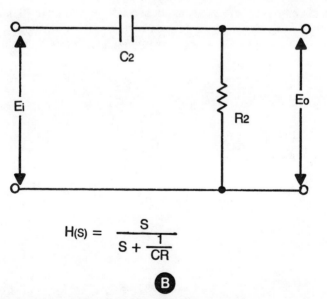

$$H_{(S)} = \frac{S}{S + \frac{1}{CR}}$$

B

Fig. 7-4. Low-pass filter transfer function when RC=0.5 seconds (A), and a high-pass filter consisting of a resistor and capacitor (B).

$$\frac{2}{S + 2}$$

This condition is preserved by dividing the constant term (in this case 40, by 2). We then supply a 2 in the numerator. The result is as follows:

$$H(S) = \frac{40}{2}\left(\frac{2}{S + 2}\right)\left(\frac{1}{S + 10}\right)$$

By the same token, the expression

$$\frac{1}{S + 10}$$

represents another low-pass filter. We then supply a 10 in the numerator and divide the constant $\frac{40}{2}$ by 10. We now have

$$H(S) = \frac{40}{2\,(10)}\left(\frac{2}{S + 2}\right)\left(\frac{10}{S + 10}\right)$$

$$H(S) = 2\left(\frac{2}{S + 2}\right)\left(\frac{10}{S + 10}\right)$$

This shows that we have a gain constant of 2, a low-pass filter with a cutoff equal to 2 radians per second, and a cutoff of 10 radians per second.

A high-pass filter as shown in Fig. 7-4 has a transfer function of

$$H(j\omega) = \frac{R}{R + \frac{1}{j\omega C}}$$

$$H(j\omega) = \frac{j\omega CR}{j\omega CR + 1}$$

$$= \frac{j\omega}{j\omega + \frac{1}{CR}}$$

If $j\omega = S$

$$H(S) = \frac{S}{S + \frac{1}{CR}}$$

This is the equation, or transfer function, for a high-pass filter using a resistor and capacitor. As an example, suppose we were given the following equation to be synthesized:

$$H(S) = \frac{60\,S}{S^2 + 12S + 20}$$

This equation could be factored as follows

$$H(S) = 60\,S\left(\frac{1}{S+2}\right)\left(\frac{1}{S+10}\right)$$

The S term in the numerator could be placed over the S + 2 term as

$$H(S) = 60\left(\frac{S}{S+2}\right)\left(\frac{1}{S+10}\right)$$

The $\frac{S}{S+2}$ term is the transfer function of a high-pass filter with a cutoff of 2 radians per second. By multiplying the term $\frac{1}{S+10}$, and dividing the constant term, which is 60, by 10, we have

$$H(S) = \frac{60}{10}\left(\frac{S}{S+2}\right)\left(\frac{10}{S+10}\right)$$

This expression now represents an amplifier with a voltage gain of 6, a low-pass filter of cutoff frequency equal to 10 radians per second, and a high-pass filter with a cutoff frequency of two radians per second. See Fig. 7-5 for the entire circuit.

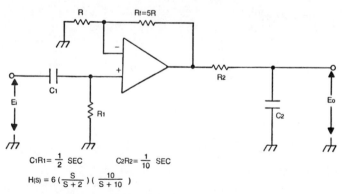

$$C_1R_1 = \frac{1}{2}\ \text{SEC} \qquad C_2R_2 = \frac{1}{10}\ \text{SEC}$$

$$H(S) = 6\left(\frac{S}{S+2}\right)\left(\frac{10}{S+10}\right)$$

Fig. 7-5. Circuit representing the function:

$$H(s) = \frac{60\,s}{(s+2)(s+10)}$$

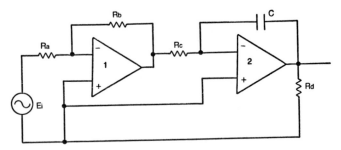

Fig. 7-6. Basic gyrator circuit.

GYRATOR

A gyrator simulates an inductor in a network. At times, a filter may require an extremely large value of inductance if a passive configuration is desired. However, there are many situations where we need the large inductance value, but the filter must be physically as small as possible. The inductance is then simulated with active components, resistors, and capacitors.

In Fig. 7-6, we see a gyrator circuit. Note that it is composed of 4 resistors, 2 operational amplifiers and 1 capacitor.

A voltage, E_i, is applied to the input of the circuit composed of op amp number 1, R_a and R_b. The output of this stage, as given by simple operational-amplifier theory, is

$$E_2 = E_1 \left(\frac{-R_b}{R_a} \right)$$

Equation 7-1

Voltage E_2 is applied to the input of the second amplifier, which has an output given by

$$E_4 = E_3 \frac{(-1)}{\frac{j \omega C}{R_c}}$$

Equation 7-2

$$E_4 = E_3 \left(-\frac{1}{j \omega C R_c} \right)$$

Since E_2 and E_3 are the same, we may write

$$E_4 = (E_1) \left(\frac{-R_b}{R_a} \right) \left(\frac{-1}{j \omega CR_c} \right)$$

Equation 7-3

166

$$= E_1 \frac{R_b}{j\omega C R_a R_c}$$

This voltage appears across resistor R_d. The current which flows through R_d is given by

$$I = \frac{E_i \left(\dfrac{R_b}{j\omega C R_a R_c} \right)}{R_d} \qquad \text{Equation 7-4}$$

$$I = \frac{E_i R_b}{j\omega C R_a R_c R_d}$$

If we divide I by E_i, we obtain

$$\frac{I}{E_i} = \frac{R_b}{j\omega C R_a R_c R_d} \qquad \text{Equation 7-5}$$

Taking the reciprocal, we obtain

$$\frac{E_i}{I} = \frac{j\omega C R_a R_c R_d}{R_b} \qquad \text{Equation 7-6}$$

Since there is a j term in the expression, and the expression denotes a ratio of voltage to current, the formula must indicate reactance. Since the reactance has a +j sign and it is proportional to frequency, the reactance must be inductive. The inductance produced will be given by

$$L = \frac{C R_a R_c R_d}{R_b} \qquad \text{Equation 7-7}$$

Another example of using a gyrator is shown in Fig. 7-7A. This is a high-pass filter composed of two high-pass sections. One section consists of a capacitor and inductor, and another section composed of a resistor and capacitor. The inductor can be replaced by the gyrator circuit as shown in Fig. 7-7B.

BAND-PASS FILTER USING A NOTCH FILTER

A band pass filter can be constructed by incorporating a notch filter in the feedback loop of an amplifier as shown in Fig. 7-8.

Basic feedback theory shows the gain for the amplifier will be

$$Av_{(0)} = \frac{A}{1 + BA} \qquad \text{Equation 7-8}$$

where A is the gain of the amplifier without feedback, and B is a feedback coefficient. If B is frequency dependent as in a band-rejection filter, we have B given by the equation

$$B = \frac{S^2 + k_1}{S^2 + Sk_2 + k_3} \qquad \text{Equation 7-9}$$

At low frequencies, a band-rejection or notch filter has a large response value. Therefore, there is considerable feedback and little gain is obtained. The same effect occurs at high frequencies. However, at the center frequency the system has maximum gain since the band rejection or notch filter has very small response, therefore little feedback can occur.

By placing resistors R_1 and R_2 into the system, we can control the voltage gain. The maximum gain of the network in Fig. 7-8 would be given by

$$Av = -\frac{R_2}{R_1} \qquad \text{Equation 7-10}$$

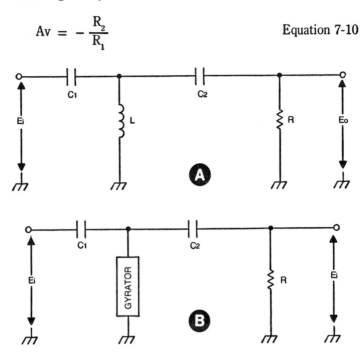

Fig. 7-7. Example of a circuit that could use a gyrator (A), and with the inductor replaced by a gyrator (B).

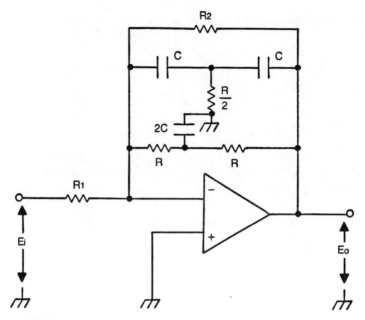

Fig. 7-8. Bandpass filter constructed by using a notch filter in the feedback loop of an amplifier.

The center frequency would be given by

$$F_o = \frac{1}{2 \pi RC}$$

Equation 7-11

Problems

1. Synthesize the following transfer functions using active devices and passive components.

a. $H(S) = \dfrac{10,000\,S}{S^2 + 3000S + 2,000,000}$

b. $H(S) = \dfrac{10,000}{S^2 + 3000S + 2,000,0000}$

c. $H(S) = \dfrac{10,000\,S^2}{S^2 + 3000S + 2,000,000}$

d. $H(S) = \dfrac{10,000\,S}{S^3 + 70\,S^2 + 1400S + 8000}$

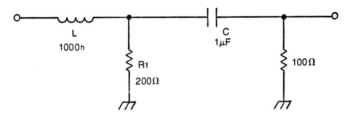

Fig. 7-9. Circuit for problem 2.

2. Convert the filter circuit shown in Fig. 7-9 to a filter containing a gyrator. Choose components to give the filter the appropriate value of inductance.

3. Draw a schematic of a circuit that can be used to produce capacitance by using active designs, inductors, and resistors.

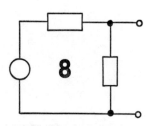

Experiments in
Synthesis and Analysis

The following are some experiments in synthesis and analysis.

BRIDGE-TEE NETWORK (EXPERIMENT 1)

Objective: To make some measurements on a bridge-tee network, and to apply some of the rules for network analysis.

Discussion: In this experiment you will use the wye to delta conversion for determining the equivalent circuit of a bridge-tee network. Then you will test the results by building the circuit.

Equipment:

    Dc voltmeter
    Dc current meter
    Resistance decade box

Procedure:

1. Determine the equivalent circuit for the bridge-tee network using the wye to delta conversion formula, see Fig. 8-1.

2. Suppose the load to be placed on the circuit is 1,000 ohms. What current do you expect in the load?

3. Build the bridge-tee network. Connect the load to the circuit. Measure the current through the load and voltage across the load, for load values between 50 ohms and 100 k-ohms. (Use a decade box to vary resistance).

4. Plot load power versus load resistance for the data measured in this experiment.

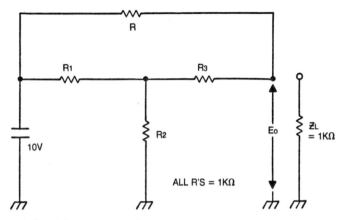

Fig. 8-1. Circuit for experiment 1.

Conclusions:

1. Under what load condition does maximum power transfer occur?

2. How did the current calculated for procedure two compare with the measured value for procedure three?

3. Derive the voltage-transfer function for a bridge-tee network as shown in Fig. 8-1. Place the values of the components in the formula and calculate the numerical value of the transfer function. Compare the answer with that calculated from the measured value of output voltage in procedure three.

4. What is the value of characteristic impedance of the network? Use this formula:

$$Z_o = \sqrt{(Z_{oc}) Z_{sc}}$$

Where: Z_o is the characteristic impedance,
Z_{oc} is the input impedance of the network when the output is open circuited.
Z_{sc} is the input impedance of the network when the output is short circuited.

CHARACTERISTIC IMPEDANCE (EXPERIMENT 2)

Objective: To determine the characteristic impedance for several networks.

Discussion: The characteristic impedance of a network may be defined as that impedance which, when placed across the output terminals of the network, causes the input impedance to become

172

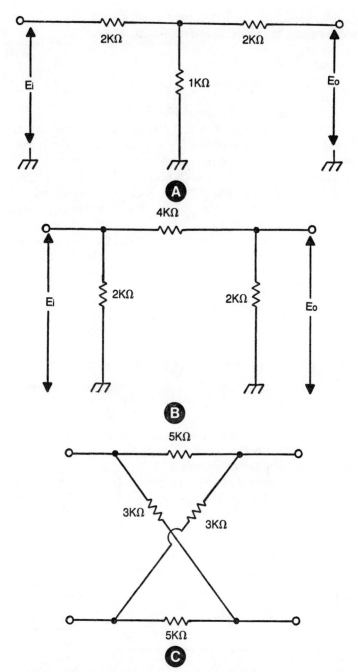

Fig. 8-2. Circuits for experiment 2.

equal to that impedance. In this experiment we will use only resistances so that the impedances are all real values.

Equipment:

Ohmmeter

Resistance decade box

Procedure:

1. Build each of the networks in Fig. 8-2.

2. Connect the ohmmeter to the input, and the decade box to the output of the circuit of Fig. 8-2A. Vary the decade box from lowest possible value to highest possible value, measuring the input resistance of the network for each decade box setting.

3. Repeat this procedure two for each circuit in Fig. 8-2.

4. Plot input resistance versus decade box resistance for each circuit in Fig. 8-2. Determine the value of decade-box resistance that causes the input resistance to have the same value as the decade box resistance. This is the characteristic impedance for the network.

Conclusions and Questions:

1. Calculate the characteristic impedance for each network and compare with the value measured in this experiment.

2. In your own words, define characteristic impedance.

Y AND Z PARAMETERS OF A NETWORK (EXPERIMENT 3)

Objective: To measure the Y and Z parameters for several different networks.

Discussion: A single-output electric network is a two-port circuit, having one output with two terminals and one input with two terminals. In other words, it has two pairs of terminals, an input or primary, and an output or secondary.

In Fig. 8-3, we see a two-port network. The currents I_1 and I_2 can be expressed in terms of the voltages V_1 and V_2 and vice versa. From the current equations, we can derive short-circuit admittances, and for the voltage equations we can derive open-circuit impedances. The defining equations for current are

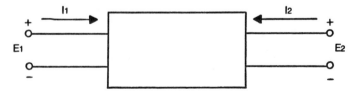

Fig. 8-3. Circuit for experiment 3.

$$I_1 = (Y_{11})(E_1) + (Y_{12})(E_2)$$

$$I_2 = (Y_{21})(E_1) + (Y_{22})(E_2)$$

The parameters are

$$Y_{11} = \left.\frac{I_1}{E_1}\right|\ E_2 = 0$$

$$Y_{21} = \left.\frac{I_2}{E_1}\right|\ E_2 = 0$$

$$Y_{12} = \left.\frac{I_1}{E_2}\right|\ E_1 = 0$$

$$Y_{22} = \left.\frac{I_2}{E_2}\right|\ E_1 = 0$$

The condition ($E_2 = 0$ or $E_1 = 0$) means that a short has been applied across the terminals where E_2 or E_1 would normally occur. The equivalent Y network is shown in Fig. 8-4. The defining equations for voltage are:

$$E_1 = (Z_{11})(I_1) + (Z_{12})(I_2)$$

$$E_2 = (Z_{21})(I_1) + (Z_{22})(I_2)$$

$$Z_{11} = \left.\frac{E_1}{I_1}\right|\ I_2 = 0$$

$$Z_{21} = \left.\frac{E_2}{I_1}\right|\ I_2 = 0$$

$$Z_{12} = \left.\frac{E_1}{I_2}\right|\ I_1 = 0$$

$$Z_{22} = \left.\frac{E_2}{I_2}\right|\ I_1 = 0$$

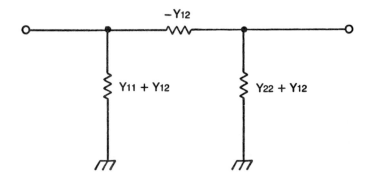

Fig. 8-4. Y-circuit for experiment 3.

The condition ($I_2 = 0$ or $I_1 = 0$) means that a open circuit exists between the terminals through which I_2 or I_1 would normally flow. The equivalent Z network is shown in Fig. 8-5.

Equipment:
  Low-voltage power supply or batteries
  Dc Ammeter
  Dc Voltmeter

Procedure:
1. Gather information to determine the Y and Z parameters for the circuits shown in Figs. 8-6A and 8-6B.

Conclusions and questions:
1. Using the information gathered from this experiment, calculate the Y and Z parameters for each network.

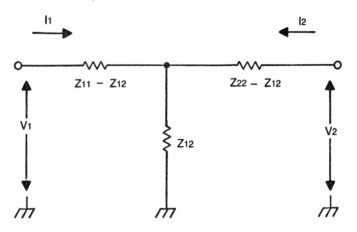

Fig. 8-5. Z-circuit for experiment 3.

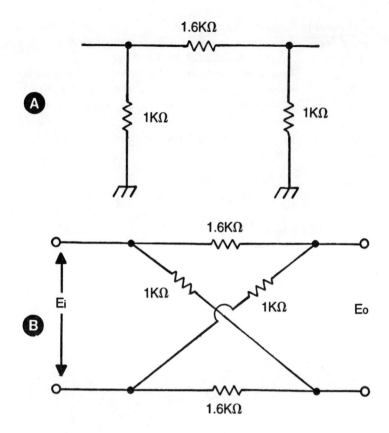

Fig. 8-6. Circuits for experiment 3 procedure.

2. For each circuit in this experiment, derive the Y and Z parameters.

EXPERIMENTAL DETERMINATION OF
ABCD CONSTANTS OF A NETWORK (EXPERIMENT 4)

Objective:

To determine the ABCD constants of several networks.

Discussion:

If a four terminal network has been constructed and can be tested, its ABCD constants may be readily determined regardless of the complexity of the network.

The A parameter represents the reciprocal of the voltage-transfer function of the network. The B parameter represents reverse transresistance of the network. Transresistance is simply the reciprocal of transconductance. The C parameter is the reverse transconductance of the network. The D parameter is the reciprocal of the current transfer function of the network.

When two networks are connected in series, and we know the ABCD parameters of the individual networks, we may use the following formulas to determine the overall parameters. Note that order is important. A_1, B_1, C_1 and D_1 belong to the first network and A_2, B_2, C_2 and D_2 belong to the second network:

$$A_T = A_1 A_2 + B_1 C_2 \qquad C_T = C_1 A_2 + D_1 C_2$$

$$B_T = A_1 B_2 + B_1 D_2 \qquad D_T = C_1 B_2 + D_1 D_2$$

In a similar manner, if networks are connected in parallel we have the following formulas:

$$A_T = \frac{A_1 B_2 + A_2 B_1}{B_1 + B_2}$$

$$B_T = \frac{B_1 B_2}{B_1 + B_2}$$

$$C_T = C_1 + C_2 + \frac{(A_1 - A_2)(D_2 - D_1)}{B_1 + B_2}$$

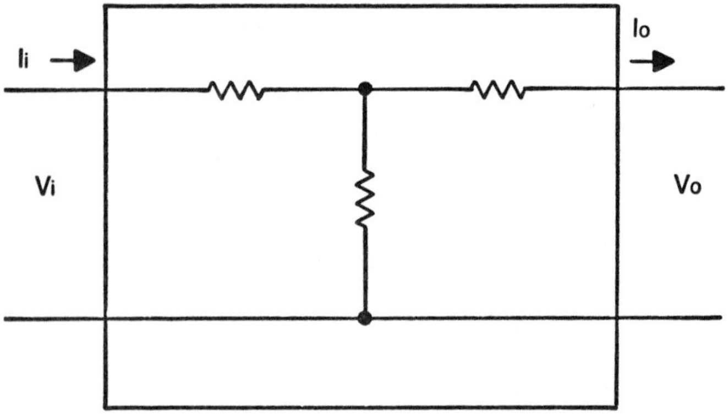

Fig. 8-7. Circuit to make measurements to determine parameters A and C in experiment 4.

$$D_T = \frac{B_2 D_1 + B_1 D_2}{B_1 + B_2}$$

Two tests are sufficient:

1. One in which the receiver terminals are open-circuited.
2. One in which the receiver terminals are short-circuited.

Test No. 1. The value of A and C may be determined from measured phasor values of V_i, I_i, and V_o as shown in Fig. 8-7. The following relations exist:

$$V_i = AV_o, \quad A = \frac{V_i}{V_o}$$

$$I_i = CV_o, \quad C = \frac{I_i}{V_o}$$

Note that the output circuit is open circuited for these measurements.

Test No. 2. The value of B and D may be determined from measured phasor values of V_i, I_o, and I_i, as shown in Fig. 8-8. The following relations exist:

$$V_i = B I_o, B = \frac{V_i}{I_o}$$

$$I_i = D I_o, D = \frac{I_i}{I_o}$$

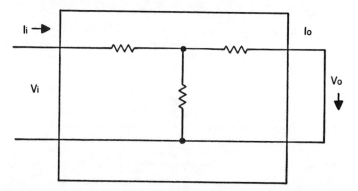

Fig. 8-8. Circuit to make measurements to determine parameters B and D in experiment 4.

179

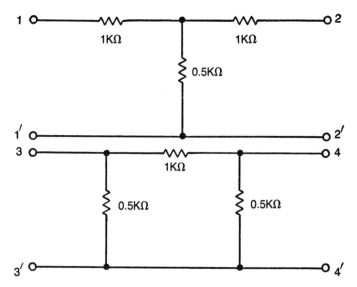

Fig. 8-9. Circuit for procedure 1 in experiment 4.

Note that the output circuit is short circuited for these measurements. In a circuit composed only of resistors, the ABCD parameters can be found by using a battery and dc meters.

Equipment:
Resistors to build circuits
VTVM, current meter
Dc power supply

Procedure:
1. Using the procedure outlined in the discussion, measure the ABCD parameters for the circuits in Fig. 8-9A and 8-9B.

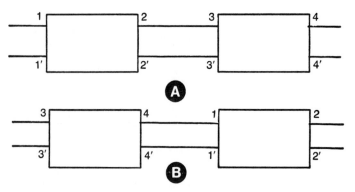

Fig. 8-10. Circuit for procedure 2 in experiment 4.

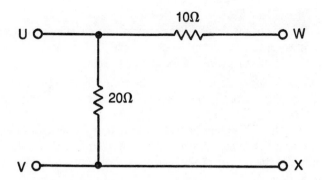

Fig. 8-11. Circuit for conclusion 4 in experiment 4.

2. Place the circuits of procedure 1 in cascade for both possible connections, as shown in Fig. 8-10A. Measure the ABCD parameters.

3. Place the circuits of procedure 1 in parallel. Measure the ABCD parameters (see Fig. 8-10B).

Conclusions and questions:

1. Calculate the ABCD parameters for the circuits of procedure 1, 2, and 3. Compare the results with the measured values.

2. Why do the overall parameters of the two networks connected in cascade differ when the order of the networks are changed?

3. Why are ABCD parameters of advantage in circuit analysis?

4. Consider the circuit shown in Fig. 8-11.

Find the ABCD parameters if
 a. UV is the input, WX is the output.
 b. WX is the input, UV is the output.

5. Compute the following for the circuits shown in Fig. 8-12.

a. ABCD parameters if the circuits are in parallel.

b. ABCD parameters if the circuits are in cascade. The circuit of Fig. 8-12A comes first.

c. ABCD parameters if the circuits are in cascade. The circuit of Fig. 8-12A comes second.

TRANSIENT RESPONSE OF A BAND-PASS NETWORK (EXPERIMENT 5)

Objective:

To become familiar with some transient measurements on low-pass filters.

Discussion:

This experiment will explore some of the characteristics of a filter in regards to transient response. We should become familiar with the following figures of merit in regards to:

Rise Time. The rise time of the step response is defined as the time required for the step response to rise from 10 to 90 percent of its final value.

Ringing. This is an oscillatory transient occurring in the response as a result of a sudden change in input (such as a step). A quantitative measure of the ringing in a step response is given by its settling time.

Settling Time. That time, ts, beyond which the step response does not differ from the final value by more than ± 2%.

Delay Time. That time, td, which the step response requires to reach 50% of its final value.

Overshoot. The difference between the peak value and the final value of the step response expressed as a percentage of the final value.

See Fig. 8-13 for an illustration of each of these quantities.

Equipment:
 Parts to build circuit
 Audio signal generator
 Dc power supply
 Oscilloscope

Procedure:

1. Build the circuit shown in Fig. 8-14.
2. Apply a square wave to the circuit. Use a frequency of 2 kHz.

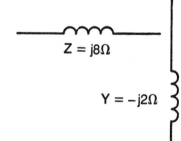

Fig. 8-12. Circuit for conclusion 5 in experiment 4.

$Z = j8\Omega$

$Y = -j2\Omega$

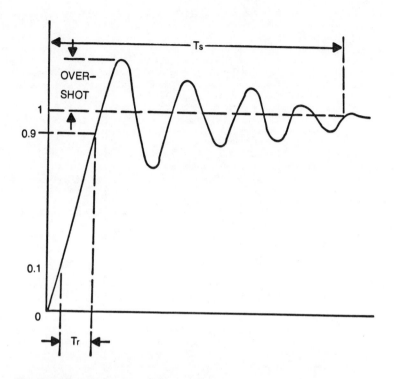

Fig. 8-13. Response diagram for experiment 5.

3. Observe the output pattern on the oscilloscope. Draw the output waveform showing:

a. ringing
b. rise time
c. delay time
d. settling time
e. overshoot

Make sure that you gather enough information to show all numerical values of the terms listed above.

Conclusions and questions:

1. Explain in detail how the circuit functions to produce a ringing waveform.

2. Calculate the Q of the tank circuit.

3. What was the ringing frequency for the tank circuit?

4. What would happen if the resistance in the tank circuit were reduced to zero?

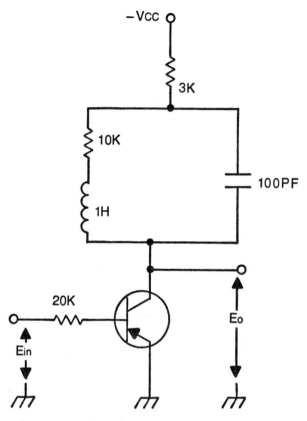

Fig. 8-14. Circuit for experiment 5.

5. What effect would you have observed had the input waveform been a sinewave instead of a square wave?

6. Write the equation for the ringing waveform that is on the output voltage for the circuit.

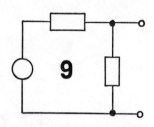

Solutions to Selected Problems and Comprehensive Exam

Chapter One

1. Let $R_1 = 30\,\Omega$, $A_2 = 10\,\Omega$,
$-jX_{c1} = -j90\,\Omega$, $-jX_{c2} = -j20\,\Omega$, $jX_{L_1} = j40\,\Omega$

$$\frac{{}^j E_o}{E_{in}} = \frac{R_2 - jX_{c2}}{R_1 + \dfrac{(-jX_{c1})\,(jX_L)}{jX_L - jX_{c1}} + R_2 - jX_{c2}}$$

$$= \frac{10 - j20}{30 + \dfrac{(-j90)\,(j40)}{j40 - j90} + 10 - j20}$$

$$= \frac{(10 - j20)\,(j40 - j90)}{30\,(j40 - j90) + 3600 + (10 - j20)\,(j40 - j90)}$$

$$= \frac{j400 - j900 + 800 - 1800}{-j1500 + 3600 - j500 - 1000}$$

$$= \frac{-j500 - 1000}{-j2000 + 2600}$$

$$= \frac{-(j500 + 1000)}{-j2000 + 2600}$$

185

$$= \frac{-(j5 + 10)}{-j20 + 26}$$

$$= \frac{-\sqrt{25 + 100} \; \angle \; \arctan \dfrac{5}{10}}{\sqrt{400 + 676} \; \angle \; \arctan \dfrac{20}{26}}$$

$$= -\frac{\sqrt{125} \; \angle \; \arctan 0.5}{\sqrt{1076} \; \angle \; -\arctan 0.77}$$

$$\frac{E_o}{E_i} = -\frac{11.2 \; \angle \; 26.5°}{32.8 \; \angle \; -37.6°} = \begin{bmatrix} -0.342 \angle 64.1 \\ 0.342 \angle -115.90° \end{bmatrix} \text{Answer}$$

2. Let Branch 1 be $Z_1 = 10 + j10$
Let Branch 2 be $Z_2 = 30 - j40$
I_1 flows in Branch 1
I_2 flows in Branch 2
By the current divider theorem,

$$I_1 = I_T \left(\frac{Z_2}{Z_1 + Z_2} \right)$$

$$= 5 \left(\frac{30 - j40}{10 + j10 + 30 - j40} \right)$$

$$= 5 \left(\frac{30 - j40}{40 - j30} \right)$$

$$= 5 \left(\frac{50 \; \angle \; -53.2°}{50 \; \angle \; -36.8°} \right)$$

$$\begin{bmatrix} I_1 = 5 \quad (1) \angle -16°.4° \end{bmatrix} \text{ Answer}$$

$$I_2 = I_T \left(\frac{Z_1}{Z_1 + Z_2} \right)$$

$$= 5 \left(\frac{10 + j10}{40 - j30} \right)$$

$$= 5 \left(\frac{14.14 \; \angle \; 45°}{50 \; \angle \; -36.8°} \right)$$

$$= 5 \quad (.2828 \angle 81.8°)$$

$$\left[I_2 = 1.414 \angle 81.8° \right] \text{ Answer}$$

3. Remove the $8 \, \Omega$ resistor. Find the voltage transfer function.

$$\text{T.F.} = \frac{-j^3}{\dfrac{6 \, (j4)}{6 + j4} - j3}$$

$$= \frac{-j^3 \, (6 + j4)}{j24 - j3 \, (6 + j4)}$$

$$= \frac{-j3 \, (6 + j4)}{j24 - j18 + 12}$$

$$= \frac{-j18 + 12}{j6 + 12}$$

The voltage across $-j3 \, \Omega$ would be 60 (T.F.) =

$$60 \left(\frac{12 - j18}{12 + j \, 6} \right)$$

The impedance looking back with the generator shorted, would be

$$Z = \frac{6 \, (j4) \, (-j3)}{(6) \, (j4) + (j4) \, (-j3) + (6) \, (-j3)}$$

$$= \frac{72}{j24 + 12 - j18} = \frac{72}{j6 + 12}$$

Thevenin's Equivalent circuit would consist of a generator of voltage equal to $\dfrac{60 \, (12 - j18)}{12 + j6}$ in series with an impedance of $\dfrac{72}{j6 + 12}$.

The total impedance with the $8 \, \Omega$ resistor will be

$$Z_T = \frac{72}{j6 + 12} + 8$$

$$= \frac{72 + j48 + 96}{j6 + 12}$$

187

$$= \frac{168 + j48}{j6 + 12}$$

The current is

$$I = \frac{E}{Z_T}$$

$$= 60 \frac{\left(\dfrac{12 - j18}{12 + j6}\right)}{\dfrac{168 + j48}{12 + j6}} = \frac{60(12 - j18)}{168 + j48}$$

$$= \frac{60 \sqrt{(12)^2 + (18)^2} \angle -\arctan \dfrac{18}{12}}{\sqrt{(168)^2 + (48)^2} \angle \arctan \dfrac{48}{168}}$$

$$= \frac{60\sqrt{144 + 324} \angle -56.4°}{\sqrt{28{,}300 + 2304} \angle 17°}$$

$$\frac{60\sqrt{468} \angle -56.4°}{\sqrt{30604} \angle 17°}$$

$$= \frac{60(21.6) \angle -56.4 - 17}{175}$$

$$I = 7.4 \angle -73.4 \text{ Answer}$$

4. Find Z:

$$Z = \frac{-j4(6 + j3)}{-j4 + 6 + j3} = \frac{-j24 + 12}{6 - j1}$$

Finding the short circuit current

$$I = \frac{E}{Z} = \frac{100}{\dfrac{-j24 + 12}{6 - j1}}$$

$$= 100\left(\frac{6 - j1}{12 - j24}\right)$$

$$= 100 \left[\frac{\sqrt{37}}{\sqrt{720}} \right] \frac{\angle \ -9.5°}{\angle \ -63.4°}$$

$$\left[I = 0.234 \angle \ 53.9° \right] \text{ Answer}$$

The current generator is in parallel with an impedance, Z.

6. $Z_L = Z^*$

$$Z^* = \frac{(4 - j4)(4 + j4)}{4 - j4 + 4 + j4}$$

$$Z^* = \frac{16 + 16}{8} = 4 \text{ ohms} \big] \text{Answer}$$

The conjugate of 4 ohms is 4 ohms.

Chapter Two

1A.

$$Z_{(s)} = \frac{3}{1 + 16S} + \frac{1 + 48S + 64S^2}{16S}$$

Divide out the expression if possible:

$$Z_{(s)} = \frac{3}{1 + 16S} + \frac{1}{16S} + 3 + 4s$$

The expression above is comparable to

$$Z_{(s)} = \frac{R_1}{1 + R_1 C_1 S} + \frac{1}{C_2 S} + R_2 + LS$$

The expression above is represented by the circuit in Fig. 9-1, where $R_1 C_1 = 16 \text{ sec}, R_1 = 3\,\Omega, C_1 = \frac{16}{R_1} = \frac{16}{3} \text{ F}, C_2 = 16\text{F}, R_2 = 3\,\Omega$

$L = 4\text{H}$

1B.

$$Z_{(s)} = \frac{19S}{1 + 42S^2} + \frac{1 + 32S^2}{4S}$$

The expression looks like a capacitor and inductor in parallel, in series with a capacitor and inductor in series.

The expression looks like the following:

$$Z_{(s)} = \frac{j\omega L_1}{1 - \omega^2 L_1 C_1} + \frac{1}{j\omega C_2} + j\omega L_2$$

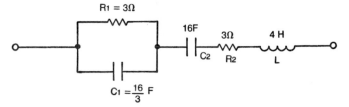

Fig. 9-1. Circuit for solution to 1A in Chapter 1.

By inspection, we have the following:

$$L_1 = 19H \qquad C_1 = \frac{L_1 C_1}{L_1} = \frac{42}{19} \text{ F} \quad C_2 = 4F, \ L_2 = 8H$$
$$L_1 C_1 = 42$$

The expression can be represented by the circuit in Fig. 9-2.

2A. $-4 + j6$ cannot be produced since it has a negative real part.

2B. $6 j\omega + \frac{1}{3} j\omega$ can be produced since it is made of a six henry coil in series with a three farad capacitor. However, we must remember that three farads of capacitance is impractical. In other words, the impedance is made of passive components both of which are not very practical.

2C. $6 j\omega + 7 j\omega^2$ cannot be produced since there is no such component that behaves like $jk\omega^2$.

3A.

$$H_{(s)} = \frac{4}{S^2 + 3S + 2} = \frac{4}{-\omega^2 + 3j\omega + 2}$$

$$|H(j\omega)| = \frac{4}{\sqrt{(2-\omega^2) + (3\omega)}} = \frac{4}{\sqrt{4 - 4\omega^2 + \omega^4 + 9\omega^2}}$$

$$= \frac{4}{\sqrt{\omega^4 + 5\omega^2 + 4}} \quad \text{(low-pass)}$$

ω	$|H(j\omega)|$	ω	$|H(j\omega)|$
0	2	4	0.22
1	1.26	5	0.14
2	0.63		
3	0.35		

See Fig. 9-3 for the magnitude response.

3B.

$$H_{(s)} = \frac{S^2 + 4}{S^2 + 3s + 2} = \frac{-\omega^2 + 4}{-\omega^2 + 3j\omega + 2}$$

190

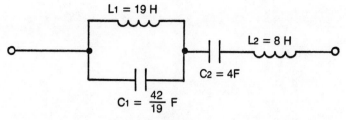

Fig. 9-2. Circuit for solution to 1B in Chapter 1.

$$|H(j\omega)| = \frac{\sqrt{(-\omega^2 + 4)^2}}{\sqrt{(2-\omega^2)^2 + (3\omega)^2}} = \frac{\sqrt{(-\omega^2 + 4)^2}}{\sqrt{\omega^4 + 5\omega^2 + 4}}$$

ω	$|H(j\omega)|$	ω	$|H(j\omega)|$	
0	2	4	0.65	
1	0.547	5	0.77	(band-
2	0			reject)
3	0.44			

See Fig. 9-4 for the magnitude response.

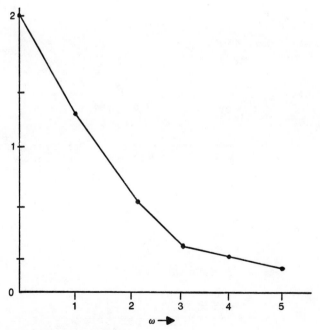

Fig. 9-3. Solution to 3A in Chapter 1.

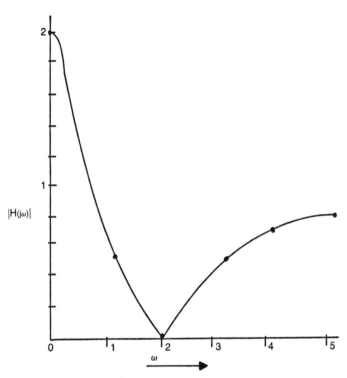

Fig. 9-4. Solution to 3B in Chapter 1.

3C.

$$H_{(s)} = \frac{4S}{S^2 + 3s + 2} = \frac{4j\omega}{-\omega^2 + 3j\omega + 2}$$

$$|H(j\omega)| = \frac{4\omega}{\sqrt{(-\omega^2 + 2)^2 + (3\omega)^2}} = \frac{4\omega}{\sqrt{\omega^4 + 5\omega^2 + 4}}$$

ω	$|H(j\omega)|$	ω	$|H(j\omega)|$
0	0	4	0.87
1	1.26	5	0.73
2	1.26		
3	1.06	high	pass

See Fig. 9-5 for the magnitude response.

3D.

$$H_{(s)} = \frac{S^2 - 3s + 2}{S^2 + 3s + 2} = \frac{-\omega^2 + 3j\omega + 2}{-\omega^2 - 3j\omega + 2}$$

$$|H(j\omega)| = \frac{\sqrt{\omega^4 + 5\omega^2 + 4}}{\sqrt{\omega^4 + 5\omega^2 + 4}} = 1 \text{ (all pass)}$$

ω	$|H(j\omega)|$	ω	$|H(j\omega)|$
0	1	4	1
1	1	5	1
2	1		
3	1		

See Fig. 9-6 for the magnitude response.

3E.

$$H_{(s)} = \frac{4^2}{S^2 + 3s + 2} = \frac{-4\omega^2}{-\omega^2 + 3j\omega + 2}$$

$$|H(j\omega)| = \frac{4\omega^2}{\sqrt{(2-\omega^2)^2 + (3\omega)^2}} = \frac{4\omega^2}{\sqrt{\omega^4 + 5\omega^2 + 4}}$$

ω	$|H(j\omega)|$	ω	$|H(j\omega)|$
0	0	4	3.48
1	1.26	5	3.64
2	2.54		
3	3.18		

See Fig. 9-7 for the magnitude response.

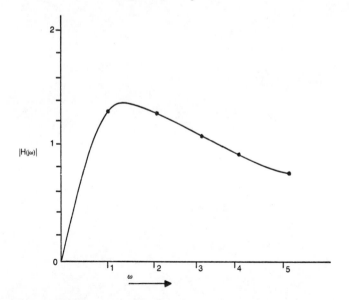

Fig. 9-5. Solution to 3C in Chapter 1.

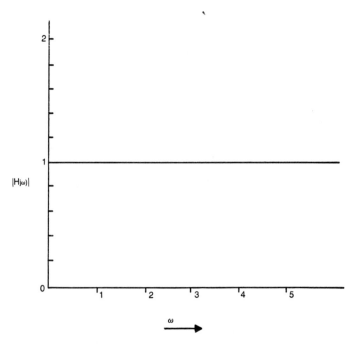

Fig. 9-6. Solution to 3D in Chapter 1.

Chapter 3

3. $H_{(s)} = \dfrac{4s + 13}{S^2 + 9s + 7}$

a. Find $H(j\omega)$

$$|H(j\omega)| = \dfrac{4j\omega + 13}{-\omega^2 + 9j\omega + 7}$$

b. Determine the phase equation

$$\theta(\omega) = \arctan \dfrac{\omega}{13} - \arctan \dfrac{9\omega}{7 - \omega^2}$$

$$- \theta(\omega) = - \arctan \dfrac{4\omega}{13} + \arctan \dfrac{9\omega}{7 - \omega^2}$$

c. $t(\omega) = \dfrac{d}{d\omega} [- \theta(\omega)]$

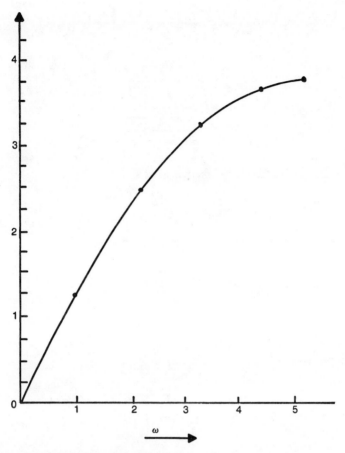

Fig. 9-7. Solution to 3E in Chapter 1.

$$t(\omega) = \frac{d}{d\omega}\left[-\arctan\frac{4\omega}{13}\right] + \frac{d}{d\omega}\left[\arctan\frac{9\omega}{7-\omega^2}\right]$$

$$\frac{d}{dx}\arctan\frac{u}{v} = \frac{1}{1+\left[\frac{u}{v}\right]^2}\ \frac{d}{dx}\left[\frac{u}{v}\right]$$

$$\frac{d}{dx}\left[-\arctan\frac{4\omega}{13}\right] = \frac{-1}{1+\frac{(4\omega)^2}{13}}\ \frac{d}{d\omega}\left[\frac{4\omega}{13}\right]$$

$$= \frac{-52}{169+16\omega^2}$$

$$\frac{d}{d} \arctan \frac{9\,\omega}{7-\omega^2} = \frac{1}{1+\dfrac{9\omega^{\,2}}{7-\omega^2}} \frac{d}{d\omega}\left[\frac{9\,\omega}{7-\omega^2}\right]$$

$$= \frac{(7-\omega^2)^2}{(7-\omega^2)^2 + (9\,\omega)^2}\left[\frac{d}{d\,\omega}\!\left(\frac{9\,\omega}{7-\omega^2}\right)\right]$$

$$= \frac{63 + 9\,\omega^2}{49 + 67\,\omega^2 + \omega^4}$$

Combining the two time equations, we have,

$$t(\omega) = \frac{-52}{160 + 16\omega^2} + \frac{63 + 9\,\omega^2}{49 + 67\omega^2 + \omega^4}$$

The following formula is very helpful in this type of problem.

$$\frac{d}{dx}\left(\frac{u}{v}\right) = \frac{v\,\dfrac{du}{dx} - u\,\dfrac{dv}{dx}}{v^2}$$

5. First find the transfer function.

$$E_i = I_1(SL + R_1) - I_2 R_1$$

$$0 = -I_1 R_1 + I_2\left(R_1 + \frac{1}{SC} + R_2\right)$$

$$\Delta = \begin{vmatrix} SL + R_1 & ,-R_1 \\ -R_1, & R_1 + \dfrac{1}{SC} + R_2 \end{vmatrix}$$

$$= (SL + R_1)\left(R_1 + \frac{1}{SC} + R_2\right) - (-R_1)^2$$

$$= S(LR_1 + LR_2) + \frac{L}{C} + R_1 R_2 + \frac{R_1}{SC}$$

$$\Delta\, I_2 = \begin{vmatrix} SL + R_1 & , E_i \\ -R_1 & , 0 \end{vmatrix} = R_1 E_i$$

$$I_2 = \frac{\Delta I_2}{\Delta} = \frac{R_1 E_i}{S(LR_1 + LR_2) + \dfrac{L}{C} + R_1 R_2 + \dfrac{R_1}{SC}}$$

$$E_o = I_2(R_2) = \left(\frac{\Delta I_2}{\Delta}\right) R_2$$

$$\frac{E_o}{E_i} = \frac{R_1 R_2 \, SC}{S^2 C \,(L_1 R_1 + LR_2) + S \,(L + C \, R_1 R_2) + R_1}$$

If $R_1 = 1$ ohm, $R_2 = 1$ ohm, $C = 1F$, $L = 1H$,

$$\frac{E_o}{E_i} = \frac{S}{2S^2 + 2S + 1}$$

Next, find the function in terms of $j\omega$:

$$\frac{E_o}{E_i} = \frac{j\omega}{-2\omega^2 + 2j\omega + 1}$$

Then find the phase equation:

$$\theta(\omega) = \frac{\omega}{0} - \arctan \frac{2\omega}{1 - 2\omega^2}$$

$$= \arctan \infty - \arctan \frac{2\omega}{1 - 2\omega^2}$$

$$t(\omega) = \frac{d\theta(\omega)}{d\omega}$$

$$= \frac{d \overbrace{\arctan \infty}^{\frac{\pi}{2}}}{d\omega} + \frac{d}{d\omega} \frac{2\omega}{1 - 2\omega^2}$$

$$= 0 + \frac{d}{d\omega} \arctan \frac{2\omega}{1 - 2\omega^2}$$

$$t(\omega) = \frac{d}{d\omega} \arctan \frac{2\omega}{1 - 2\omega^2}$$

$$= \frac{1}{1 + \left(\frac{2\omega}{1 - 2\omega^2}\right)^2} \frac{d}{d\omega} \left[\frac{2\omega}{1 - 2\omega^2}\right]$$

$$= \frac{1}{1 + \left(\frac{\omega}{1 - 2\omega^2}\right)^2} \left[\frac{(1 - 2\omega^2)^2 \frac{d \, 2\omega}{d\omega} - 2\omega \frac{d}{d\omega} 1 - 2\omega^2)}{(1 - 2\omega^2)^2}\right]$$

$$= \frac{1}{(1 - 2\omega^2)^2 + (2\omega)^2} \left[(1 - 2\omega^2)(2) - 2\omega\left[0 - 2\omega\right]\right]$$

197

$$= \frac{1}{1 - 4\omega^2 + 4\omega^4 + 4\omega^2} \left[2 - 4\omega^2 + 8\omega^2 \right]$$

$$= \frac{2 + 4\omega^2}{1 + 4\omega^4}$$

6. Find the mesh equations:

$$E_i = I_1 R + I_1 SL - I_2 SL$$
$$O = -I_1 SL + I_2 SL + I_2 R + I_2 SL$$
$$E_i = I_1 (R + SL) - I_2 SL$$
$$O = -I_1 SL + I_2 (2SL + R)$$

$$\Delta = \begin{vmatrix} R + SL, & -SL \\ -SL, & 2SL + R \end{vmatrix} = 3 SLR + R^2 + S^2L^2$$

$$\Delta I_2 = \begin{vmatrix} R + SL, & E_i \\ -SL, & O \end{vmatrix} = SL E_i$$

$$I_2 = \frac{\Delta I_2}{\Delta} = \frac{SL E_i}{S^2L^2 + 3SLR + R^2}$$

$$E_o = I_2 (SL) = \frac{S^2L^2 E_i}{S^2L^2 + 3SLR + R^2}$$

$$\frac{E_o}{E_i} = \frac{S^2}{S^2 + \frac{3SR}{L} + \frac{R^2}{L^2}}$$

$$\frac{E_o}{E_i} = \frac{S^2}{S^2 + 15S + 25}$$

7a.

$$H_{(s)} = \frac{0.01S^2 + 0.2S + 1}{0.01S^2 + 0.3S + 1}$$

$$= \frac{(RC)^2S^2 + 2S(RC) + 1}{(RC)^2S^2 + 3S(RC) + 1}$$

By inspection, we see that this equation has the following component values: $(RC)^2 = 0.01$, $RC = 0.1$ second.

If $R = 1000$ ohms, $C = \dfrac{RC}{R} = \dfrac{0.1}{1 \times 10^3} = 0.1$ mF

The circuit represented is shown in Fig. 9-8.
7b.

$$H_{(s)} = \frac{S^2 + 40,000}{S^2 + 800S + 40,000}$$

$$= \frac{S^2 + \left(\frac{1}{RC}\right)^2}{S^2 + 4\left(\frac{1}{RC}\right)S + \left(\frac{1}{RC}\right)^2}$$

By inspection:

$$40,000 = \left(\frac{1}{RC}\right)^2,$$

$$\frac{1}{RC} = 200 \text{ rad/sec}.$$

$$RC = \frac{1}{200} = 5 \times 10^{-3} \text{ sec}$$

If R = 1000 ohms,

$$C = \frac{RC}{R} = \frac{5 \times 10^{-3}}{10^3} = 5\,\mu F$$

$$\frac{R}{2} = 500\,\Omega$$

$$2C = 10\,\mu F$$

The circuit is shown in Fig. 9-9.

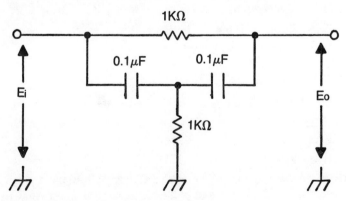

Fig. 9-8. Circuit for solution to 7A in Chapter 3.

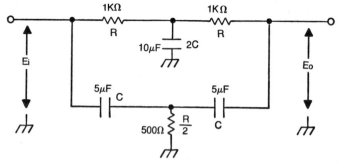

Fig. 9-9. Circuit for solution to 7B in Chapter 3.

8. $H(s) = \dfrac{S^4 + 200S^2 + 10,000}{S^4 + 80S^3 + 1800S^2 + 8000S + 10,000}$

We notice that $S^4 + 200S^2 + 10,000$ is a perfect square $(S^2 + 100)^2$. $S^2 + 100$ looks like part of the transfer function for a twin-tee band-pass filter.

We then see if we can factor the denominator. We know that if we have a quadratic equation of the form $S^2 + 2aS + b$, the square of the quadratic is:

$$S^4 + 4aS^3 + S^2 [2b + 4a^2] + 4aSb + b^2$$

Perhaps $H(s)$ denominator is a perfect square. If it were, then by inspection,

$80 = 4a, a = 20, b = 100$
$2b + 4a^2 = 200 + 4(20)^2 = 1800$
$4ab = 4(20) (100) = 8000$

Therefore, it appears as a square of

$$S^2 + 2(20)S + 100 = S^2 + 40S + 100$$

The transfer function can be written:

$$H(s) = \left(\frac{S^2 + 100}{S^2 + 40S + 100} \right)^2$$

The circuit is shown in Fig. 9-10.

24a. The output is the same as the input voltage. The transfer function is unity:

$$H(s) = 1$$

24b. Since there is no path for current through a load, there is no voltage drop in R_2. The output is the same value as is across R_1, which is the input voltage.

$$H(s) = 1$$

25.

$$\frac{E_o}{E_i} = H_{(s)} \frac{R_3}{R_1 + R_3}$$

Chapter Four

1A. Find the resonant frequency of the circuit as it is in Fig. 4-12.

$$f_r = \frac{1}{2\pi\sqrt{LC}} = \frac{0.159}{\sqrt{2\times10^{-6}\times30\times10^{-6}}} = 20\text{kHz}$$

1B. Find the scaling constant:

$$\eta = \frac{f_{new}}{f_{old}} = \frac{8000}{20,000} = 0.4$$

1C. Compute the new components:

$$L_{new} = \frac{L_{old}}{\eta} = \frac{2\,\mu\text{H}}{0.4} = 5\text{H}$$

$$C_{new} = \frac{C_{old}}{\eta} = \frac{30\,\mu\text{F}}{0.4} = 75\,\mu\text{F}$$

The value of the resistor did not change in this scaling procedure.

2A. Determine the impedance scaling constant:

$$K_z = \frac{\text{new impedance}}{\text{old impedance}} = \frac{1}{2}$$

2B. Compute the new component values.

$$R1 \text{ new} = R1 \text{ old } (K_z) = 10\left(\frac{1}{2}\right) = 5 \text{ ohms}$$

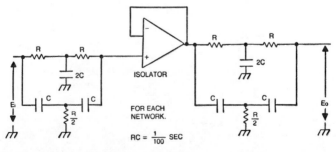

Fig. 9-10. Circuit for solution to 8 in Chapter 3.

$$R2 \text{ new} = R2 \text{ old } (K_z) = 20 \left(\frac{1}{2}\right) = 10 \text{ ohms}$$

$$L_{new} = L_{old} (K_z) = 3mH \left(\frac{1}{2}\right) = 1.5 \text{ mH}$$

$$C_{new} = \frac{C_{old}}{K_z} = \frac{4 \,\mu F}{\frac{1}{2}} = 8 \,\mu F$$

3A. First find the frequency scaling constant $K_f n$:

$$\eta = \frac{\text{new cutoff}}{\text{old cutoff}} = \frac{40,000 Hz}{.159 Hz} \cong 250,000$$

$$C_{new} = 1 \,\mu F$$

$$L_{new} = \frac{L(Kz)}{n} \,, \quad C_{new} = \frac{C_{old}}{Kz(n)} \,, \quad Kz = 8$$

$$L_{new} = \frac{1 \,(8)}{2.5 \times 10^5} = 3.2 \times 10^{-5} = 32 \,\mu H$$

$$R_{new} = R_{old} \,(Kz) = 1 \,(8) = 8 \text{ ohms.}$$

Chapter Five

1. $|H(j\omega)| = \dfrac{1}{\sqrt{1 + \left(\frac{f}{f_c}\right)^b}}$

$$= \frac{1}{\sqrt{1 + \left(\frac{10^4}{10^2}\right)^6}}$$

$$= \frac{1}{\sqrt{+ (10^2)^6}} \cong \frac{1}{10^6} = 10^{-6}$$

2.
$$|H(j\omega)| = \frac{4200}{\sqrt{1 + \omega^2}} \,,$$

$$|H(j\omega)|^2 = \frac{(4200)^2}{1 + \omega^2}$$

$$H(s) \, H(-s) = \frac{(4200)^2}{1 - S^2}$$

$$H(s) = \frac{4200}{S + 1}$$

202

$$H(j\omega) = \frac{4200}{j\omega + 1}$$

3. Proceed in the manner shown in chapter 5. The answer should be

$$H(s) = \frac{1}{S^5 + 3.24S^4 + 5.24S^3 + 5.24S^2 + 3.24S + 1}$$

4. To plot the magnitude, use the equation

$$|H(j\omega)| = \frac{1}{\sqrt{1 + (\frac{f}{fc})^8}}$$

To plot the phase, use the equation

$$\theta_{(\omega)} = -\arctan \frac{2.614\,\omega - 2.614\omega^3}{\omega^4 - 3.416\omega^2 + 1}$$

5. $|H(j\omega)| = \dfrac{1}{\sqrt{1 + E^2(Cn)^2}}$

$$= \frac{1}{\sqrt{1 + (.6)^2 \, (8\omega^4 - 8\omega^2 + 1)^2}}$$

$$= \frac{1}{\sqrt{1 + (.36)(64\omega^8 - 128\omega^6 + 80\omega^4 - 16\omega^2 + 1)}}$$

$$= \frac{1}{\sqrt{1.36 + 23\omega^8 - 46\omega^6 + 28.8\omega^4 - 5.76\,\omega^2}}$$

$$= \frac{1}{\sqrt{23\omega^8 - 46\,\omega^6 + 28.8\omega^4 - 5.76\,\omega^2 + 1.36}}$$

Chapter Six

To find Cauer I for problem 1a, we do the following:

$$
\begin{array}{r}
S \\[2pt]
S^3 + 3S \overline{\smash{\big)}\, S^4 + 6S^2 + 8} \\
\underline{S^4 + 3S^2} \\
3S^2 + 8
\end{array}
$$

$$
\begin{array}{r}
\frac{S}{3} \\[2pt]
3S^2 + 8 \overline{\smash{\big)}\, S^3 + 3S} \\
\underline{S^3 + 2.67S} \\
0.33S
\end{array}
$$

$$0.33S \overline{\smash{\big)}\, 3S^2 + 8} \quad \frac{9.09\,S}{}$$

$$\underline{3S^2}$$
$$8$$

$$8 \overline{\smash{\big)}\, 0.33S} \quad \frac{.04125\,S}{}$$
$$\underline{0.33S}$$

For the equation of Cauer I, we obtain

$$\alpha_1 = 1, \ \alpha_2 = \frac{1}{3}, \ \alpha_3 = 9.09, \ \alpha_4 = .04125$$

The resulting circuit is shown in Fig. 9-11.

Check of Cauer I for problem 1a:

$$Z_{(S)} = SL_1 \quad \frac{\dfrac{1}{C_1 S}\left(SL_2 + \dfrac{1}{C_2 S}\right)}{\dfrac{1}{C_1 S} + SL_2 + \dfrac{1}{C_2 S}}$$

$$= S + \frac{\dfrac{3}{S}\left(9.095 + \dfrac{24}{S}\right)}{\dfrac{27}{S} + 9.09S}$$

$$= S + \frac{3\left(9.09S + \dfrac{24}{S}\right)}{27 + 9.09\,S^2}$$

$$= \frac{27\,S + 9.09\,S^3 + 27.27\,S + \dfrac{72}{S}}{27 + 9.09\,S^2}$$

$$= \frac{27\,S^2 + 9.09\,S^4 + 27.27\,S^2 + 72}{27\,S + 9.09\,S^3}$$

$$= \frac{9.09\,S^4 + 54.27\,S^2 + 72}{9.09\,S^3 + 27\,S}$$

$$\simeq \frac{9.09\,S^4 + 54.27\,S^2 + 72}{(9.09)\,(S^3 + 3S)}$$

$$= \frac{S^4 + 6\,S^2 + 8}{S^3 + 3\,S}$$

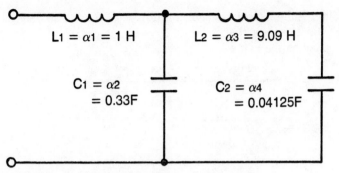

Fig. 9-11. Circuit for solution to 1A (Cauer I) in Chapter 6.

To find Cauer II for problem 1b, we do the following:

$$
\begin{array}{r}
\dfrac{8}{3S} \\[4pt]
3\,S + S^3 \overline{\smash{\big)}\,8 + 6\,S^2 + S^4} \\[2pt]
\underline{8 + 2.67\,S^2} \\[2pt]
5.33S^2 + S^4
\end{array}
$$

$$
\begin{array}{r}
\dfrac{0.56/S}{} \\[4pt]
5.33S^2 + S^4 \overline{\smash{\big)}\,3S + S^3} \\[2pt]
\underline{3S + .56S^3} \\[2pt]
.44\,S^3
\end{array}
$$

$$
\begin{array}{r}
\dfrac{12.1}{S} \\[4pt]
0.44S^3 \overline{\smash{\big)}\,5.33S^2 + S^4} \\[2pt]
\underline{5.33S^2} \\[2pt]
S^4
\end{array}
$$

$$
\begin{array}{r}
\dfrac{0.44}{S} \\[4pt]
S^4 \overline{\smash{\big)}\,0.44S^3} \\[2pt]
\underline{0.44S^3}
\end{array}
$$

For the equation of Cauer II, we obtain

$$
b_1 = \frac{3}{8}\ \text{F},\quad b_2 = \frac{1}{0.56} = 1.785\ \text{H},\quad b_3 = \frac{1}{12.1} = .0826\text{F}
$$

$$
b_4 = \frac{1}{0.44} = 2.27\ \text{H}
$$

The resulting circuit is shown in Fig. 9-12. Let's also determine Foster I & II for this problem.

We obtain the following:

1a. $Z_{(S)} = \dfrac{(S^2 + 2)(S^2 + 4)}{S(S^2 + 3)} = \dfrac{S^4 + 6S^2 + 8}{S^3 + 3S}$

$$
\begin{array}{r}
S \\
S^3 + 3S \overline{\smash{\big)}\ S^4 + 6S^2 + 8} \\
\underline{S^4 + 3S^3} \\
3S^2 + 8
\end{array}
$$

$Z_{(S)} = S + \dfrac{3S^2 + 8}{S^3 + 3S}$

$\dfrac{3S^2 + 8}{S^3 + 3S} = \dfrac{3S^2 + 8}{S(S^2 + 3)} = \dfrac{A}{S} + \dfrac{BS}{S^2 + 3}$

Multiply both sides of equation by $S(S^2 + 3)$:

$3S^2 + 8 = A(S^2 + 3) + BS^2$

Let $S^2 = 0$

$8 = A(3)$, $A = \dfrac{8}{3}$

Let $S^2 = -3$

$3(-3) + 8 = B(-3)$

$-1 = B(-3)$

$B = \dfrac{1}{3}$

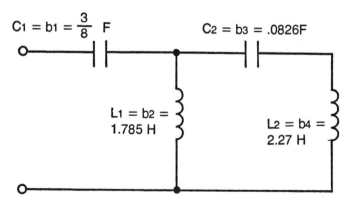

$C_1 = b_1 = \dfrac{3}{8}$ F $C_2 = b_3 = .0826F$

$L_1 = b_2 = 1.785$ H $L_2 = b_4 = 2.27$ H

Fig. 9-12. Circuit for Solution to 1A (Cauer II) in Chapter 6.

$$Z_{(S)} = S + \frac{\frac{8}{3}}{S} + \frac{\frac{1}{3}S}{S^2 + 3}$$

By comparison to the Foster I equation,

$$Z_{(S)} = K_a S + \frac{K_b}{S} + \frac{2 K_c S}{S^2 + \omega_c^2}$$

We obtain the following:

$$K_a = 1 \ , \ K_b = \frac{8}{3} \ , \ 2K_c = \frac{1}{3} \ , \ \omega_c^2 = 3$$

$$L_1 = K_a = 1 \ H$$

$$C_1 = \frac{1}{K_b} = \frac{3}{8} \ F$$

$$C_2 = \frac{1}{2 K_c} = \frac{1}{\frac{1}{3}} = 3 \ F$$

$$L_2 = \frac{2 K_c}{\omega_c^2} = \frac{\frac{1}{3}}{3} = \frac{1}{9} \ H$$

The complete circuit is shown in Fig. 9-13.
To find Foster's second approximation, find $Y_{(S)}$.

$$Y_{(S)} = \frac{1}{Z_{(S)}} = \frac{S (S^2 + 3)}{S^4 + 6S^2 + 8}$$

Factor, and set up partial fractions;

$$Y_{(S)} = \frac{S (S^2 + 3)}{(S^2 + 4) (S^2 + 2)} = \frac{A_S}{S^2 + 4} + \frac{B_S}{S^2 + 2}$$

Multiply through by $(S^2 + 4) (S^2 + 2)$:

$$S (S^2 + 3) = A_S (S^2 + 2) + B_S (S^2 + 4)$$

$$S^2 + 3 = A (S^2 + 2) + B (S^2 + 4)$$

Let $S^2 = -2$,

$$-2 + 3 = B (-2 + 4) \ , \ B = \frac{1}{2}$$

Let $S^2 = -4$

$$-4 + 3 = A (-4 + 2) \ , \ A = \frac{1}{2}$$

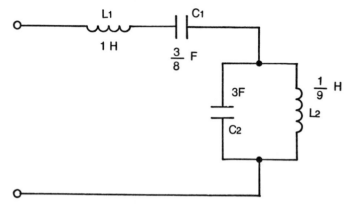

Fig. 9-13. Circuit for solution to 1A (Foster I) in Chapter 6.

Check of Problem 1a:

Foster I: Write the impedance equation for the circuit in Fig. 9-13.

$$Z_{(S)} = SL_1 + \frac{1}{C_1 S} + \frac{(SL_2) \dfrac{1}{C_2 S}}{SL_2 + \dfrac{1}{C_2 S}}$$

$$= S_{(1)} + \frac{1}{\left(\dfrac{3}{8}\right)S} + \frac{\left(S\left(\dfrac{1}{9}\right)\right)\left(\dfrac{1}{3S}\right)}{S\left(\dfrac{1}{9}\right) + \dfrac{1}{3S}}$$

$$= S + \frac{8}{3S} + \frac{\dfrac{1}{27}}{\dfrac{S}{9} + \dfrac{1}{3S}} \qquad = S + \frac{8}{3S} + \frac{\dfrac{1}{27}}{\dfrac{3S^2 + 9}{27\,S}}$$

$$= S + \frac{8}{3S} + \frac{S}{3S + 9}$$

$$= \frac{3\,S^2 + 8}{3S} + \frac{S}{3S^2 + 9}$$

$$= \frac{(3S^2 + 8)\ (3S^2+9)\ + 3S^2}{3S\qquad (3S^2 + 9)}$$

$$= \frac{9S^4 + 54\,S^2 + 72}{3S\,(3S^2 + 9)}$$

$$= \frac{9S^4 + 54S^2 + 72}{9S^3 + 27S}$$

$$= \frac{S^4 + 6S^2 + 8}{S\,(S^2 + 3)}$$

The answer agrees with the original impedance equation.
Foster II: Write the impedance equation for the circuit in Fig. 9-14.

$$Z_{(S)} = \frac{\left(SL_1 + \frac{1}{C_1 S}\right)\left(SL_2 + \frac{1}{C_2 S}\right)}{SL_1 + \frac{1}{C_1 S} + \frac{1}{C_2 S} + SL_2}$$

$$Z_{(S)} = \frac{\left(2S + \frac{8}{S}\right)\left(2S + \frac{4}{S}\right)}{2S + \frac{12}{S} + 2S}$$

$$= \frac{4\,S^2 + 24 + \frac{32}{S^2}}{4\,S + \frac{12}{S}}$$

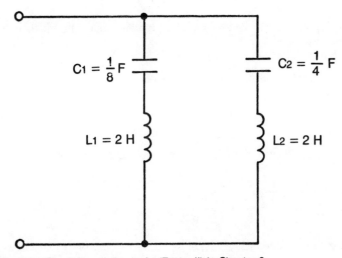

Fig. 9-14. Circuit for solution to 1a (Foster II) in Chapter 6.

$$= \frac{\dfrac{4\,S^4 + 24\,S^2 + 32}{S^2}}{\dfrac{4\,S^2 + 12}{S}}$$

$$= \frac{4\,S^4 + 24\,S^2 + 32}{(4\,S^2 + 12)\,S}$$

$$= \frac{S^4 + 6\,S^2 + 8}{S\,(S^2 + 3)}$$

The answer agrees with the original impedance equation.

3b. $Z_{(S)} = 16 + \dfrac{9S}{4 + 3S}$

$$= 16 + \frac{\dfrac{9S}{9S}}{\dfrac{4}{9S} + \dfrac{3S}{9S}}$$

$$= 16 + \frac{1}{\dfrac{1}{\dfrac{9}{4}S} + \dfrac{1}{3}}$$

The 16 represents a 16-ohm resistor which is in a series with a parallel circuit having an admittance of $\dfrac{1}{3} + \dfrac{1}{\dfrac{9}{4}S}$.

The $\dfrac{1}{3}$ is the conductance of the circuit component which is a 3-ohm resistor.

The $\dfrac{1}{\dfrac{9}{4}S}$ is the inductive susceptance of the parallel circuit with a $\dfrac{9}{4}$ henry coil (2.25 H).

The circuit is shown in Fig. 9-15.

Write the equation for $Y_{(S)}$ using the values of A and B:

$$Y_{(S)} = \frac{\dfrac{1}{2}S}{S^2 + 4} + \frac{\dfrac{1}{2}S}{S^2 + 2}$$

Comparing the admittance equation to Foster's second equation, we have.

210

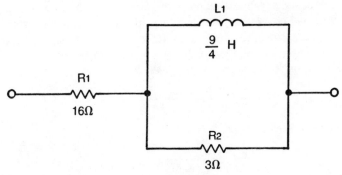

Fig. 9-15. Circuit for solution to 3B in Chapter 6.

$$Y_{(S)} = K'_a S + \frac{K'_b}{S} + \frac{2 K'_c S}{S^2 + \omega_c^2} + \frac{2 K'_d S}{S^2 + \omega_d^2}$$

where $K'_a = 0$, $K'_b = 0$, $2 K'_c = \frac{1}{2}$, $2 k'_d = \frac{1}{2}$ $\omega^2_c = 4$, $\omega^2 d = 2$.

$$L_1 = \frac{1}{2K'_c} = 2H$$

$$C_1 = \frac{2K'_c}{\omega_c^2} = \frac{1}{8} F$$

$$L_2 = \frac{1}{2K'_d} = 2H$$

$$C_2 = \frac{2K'_d}{\omega_d^2} = \frac{1}{-} F$$

The complete circuit is shown in Fig. 9-14.

Chapter Seven

1a. The transfer function was given as

$$H_{(S)} = \frac{10,000\, S}{S^2 + 3000S + 2,000,000}$$

Factor the expression as follows:

$$H_{(S)} = \frac{10,000\, S}{(S + 2000)\,(S + 1000)}$$

$$= S\, \frac{S}{S + 1000}\, \frac{2000}{S + 2000}$$

This equation represents an amplifier with a gain of five, used as isolation between a low-pass filter with a cutoff frequency of 2000 radians/second and a high-pass filter with a cutoff frequency of 1000 radians/second. See Fig. 9-16 for the schematic.

1b. The transfer function was given as

$$H_{(s)} = \frac{10,000}{S^2 + 3000\,S + 2,000,000}$$

Factor the expression as follows:

$$H_{(S)} = \frac{10,000}{(S + 2000)\,(S + 1000)}$$

$$H_{(S)} = \frac{10,000}{(2000)\,(1000)}\left(\frac{2000}{S + 2000} \times \frac{1000}{S + 1000}\right)$$

$$= \frac{10^4}{2 \times 10^6}\left(\frac{2000}{S + 2000} \times \frac{1000}{S + 1000}\right)$$

$$= 0.5 \times 10^{-2}\left(\frac{2000}{S + 2000} \times \frac{1000}{S + 1000}\right)$$

$$= 5 \times 10^{-3}\left(\frac{2000}{S + 2000}\right)\left(\frac{1000}{S + 1000}\right)$$

This equation represents two low-pass filters, of cutoff frequencies equal to 2000 and 1000 radians/second respectively, isolated by a gain of 5×10^{-3}.

See Fig. 9-17 for the schematic.

1c. The transfer function was given as

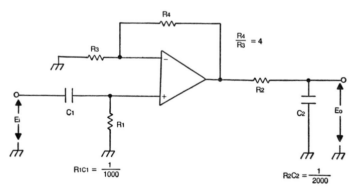

$$\frac{R_4}{R_3} = 4$$

$$R_1C_1 = \frac{1}{1000}$$

$$R_2C_2 = \frac{1}{2000}$$

Fig. 9-16. Circuit for solution to 1A in Chapter 7.

212

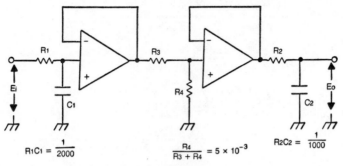

$$R_1C_1 = \frac{1}{2000}$$

$$\frac{R_4}{R_3 + R_4} = 5 \times 10^{-3}$$

$$R_2C_2 = \frac{1}{1000}$$

Fig. 9-17. Circuit for solution to 1B in Chapter 7.

$$H_{(S)} = \frac{10,000\ S^2}{S^2 + 3000\ S + 2,000,000}$$

Factor the expression as follows:

$$H_{(s)} = 10,000 \left(\frac{S}{S + 1000}\right)\left(\frac{S}{S + 2000}\right)$$

This expression represents an amplifier of gain equal to 10,000, isolating two filters of cut off frequencies equal to 1000 radians/second and 2000 radians/second. The filters are both high-pass types. See Fig. 9-18 for the schematic.

1d. The transfer function was given as

$$H_{(s)} = \frac{10,000\ S}{S^3 + 70\ S^2 + 1400\ S + 8000}$$

Factor the transfer function into the following:

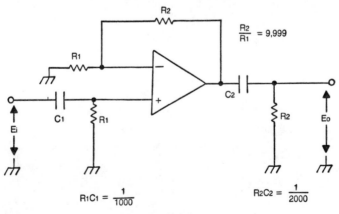

$$\frac{R_2}{R_1} = 9,999$$

$$R_1C_1 = \frac{1}{1000}$$

$$R_2C_2 = \frac{1}{2000}$$

Fig. 9-18. Circuit for solution to 1C in Chapter 7.

$$H_{(s)} = \frac{10,000\ S}{(S^2 + 30\ S + 200)\ (S + 40)}$$

$$H_{(s)} = \frac{10,000\ S}{(S + 10)\ (S + 20)\ (S + 40)}$$

$$= \left(\frac{10,000}{(20)\ (40)}\right)\left(\frac{S}{S + 10}\right)\left(\frac{20}{S + 20}\right)\left(\frac{40}{S + 40}\right)$$

$$= \left(12.5\right)\left(\frac{S}{S + 10}\right)\left(\frac{20}{S + 20}\right)\left(\frac{40}{S + 40}\right)$$

This equation represents three filters: one high-pass of cutoff equal to 10 radians/second and two low-pass filters with cutoff frequencies of 20 and 40 radians/second. Each filter is isolated from each other by an isolation stage. The total gain of the isolators is 12.5. See Fig. 9-19 for the circuit.

2. To obtain the solution to problem 7-2, simply draw the schematic of the gyrator in place of the inductor. See Fig. 9-20 for the answer.

3. To obtain a schematic of a circuit that will produce capacitance, use the original circuit for the gyrator, and replace the capacitor with an inductor. The mathematics for the circuit is shown below. The input voltage is applied between terminals A and B. The output of the first stage is given by

$$E_{o1} = (E_i)\ \left(\frac{-R_2}{R_1}\right)$$

The output of the second stage, which develops across R_4 is

$$E_{o2} = E_{o1}\ \left(\frac{-j\omega L}{R_3}\right)$$

$$= (E_i)\ \left(\frac{-R_2}{R_1}\right)\left(-\frac{j\omega L}{R_3}\right)$$

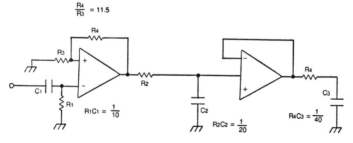

Fig. 9-19. Circuit for solution to 1D in Chapter 7.

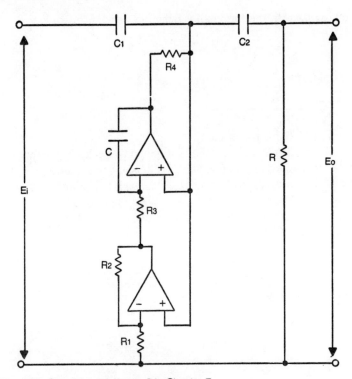

Fig. 9-20. Circuit for solution to 2 in Chapter 7.

$$E_{o2} = (E_i) \left(\frac{j\omega L \times R_2}{R_1 \times R_3} \right)$$

Current which flows through R_4 is

$$I = \frac{E_{o2}}{R_4} = \frac{(E_i)}{R_4} \left(\frac{j\omega L \times R_2}{R_1 \times R_3} \right)$$

Divide I by E_i and obtain

$$\frac{I}{E_i} = \frac{j\omega L \times R_2}{R_1 \times R_3 \times R_4}$$

Taking reciprocals,

$$\frac{E_i}{I} = \frac{R_1 \times R_3 \times R_4}{j\omega L \times R_2}$$

The $+j$ in the denominator of the impedance indicates capacitive reactance.

Comparing with the formula for inductive reactance, we have:

$$- j X_c = \frac{1}{j \omega C} = \frac{R_1 \times R_3 \times R_4}{j \omega L \times R_2}$$

It is therefore easy to see that

$$C = \frac{L \times R_2}{R_1 \times R_3 \times R_4}$$

See Fig. 9-21 for the circuit.

Now that you have finished studying this book, try the following exam in the subject of network synthesis. You should be able to complete the exam in three hours. Once you have finished, check your work against the solutions following the exam.

1. Identify each of the following as either all pass, low pass, high pass, band pass, or band reject.

a. $H_{(s)} = \dfrac{S}{S + 4}$

b. $H_{(s)} = \dfrac{10\,S}{S^2 + 19\,S + 30}$

c. $H_{(s)} = \dfrac{S^2 + 8}{S^2 + 4\,S + 8}$

d. $H_{(s)} = \dfrac{S^2}{S^2 + 3\,S + 7}$

e. $H_{(s)} = \dfrac{14}{S^2 + 19\,S + 20}$

f. $H_{(s)} = \dfrac{19}{S + 19}$

g. $H_{(s)} = \dfrac{S + 14}{S - 14}$

h. $H_{(s)} = \dfrac{S^2 + 4S + 13}{S^2 - 4S + 13}$

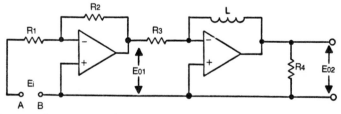

Fig. 9-21. Circuit for solution to 3 in Chapter 7.

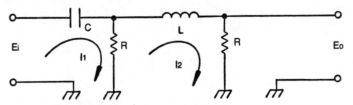

Fig. 9-22. Circuit for question 3 in Chapter 9.

2. Given the following transfer function:

$$H_{(s)} = \frac{150\ S}{S^2 + 8\ S + 12}$$

a. Find $H_{(j\omega)}$

b. Find $|\ H_{(j\omega)}\ |$

c. Find $\theta_{(\omega)}$

3. Derive the transfer function for the circuit in Fig. 9-22.

4. Given the following transfer function:

$$H_{(j\omega)} = \frac{1}{1 + j\ 2\omega}$$

Determine the value of $t_{(\omega)}$ at a frequency of 3 radians per second.

5. Write the transfer function for the circuit in Fig. 9-23.

6. Determine the value of the transfer function for the circuit in Fig. 9-24 at a frequency of $\frac{R}{L} = \omega_o$.

7. Scale the circuit of Fig. 9-25 so that it has a cutoff frequency of 5000 Hz. The cutoff frequency of the circuit is now 1000 Hz.

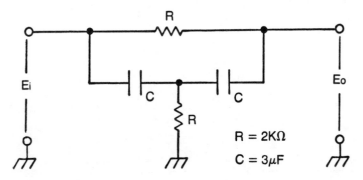

Fig. 9-23. Circuit for question 5 in Chapter 9.

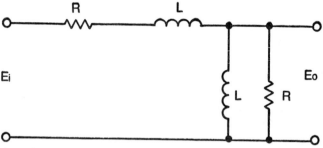

Fig. 9-24. Circuit for question 6 in Chapter 9.

8. Describe the differences in the magnitude curves for a Butterworth and Chebyshev filter.

9. Synthesize the following by any method covered in this book:

a. $Z_{(s)} = \dfrac{3 S^4 + 9 S^2 + 3}{S^3 + 2S}$

b. $Z_{(s)} = \dfrac{S^4 + 12 S^2 + 27}{S (S^2 + 6)}$

c. $Z_{(s)} = \dfrac{19 (S^2 + 8)}{S (S^2 + 10)}$

10. Synthesize the following equation using Foster II:

a. $Z_{(s)} = \dfrac{3 S^4 + 9S^2 + 3}{S^3 + 2S}$

b. $Z_{(s)} = \dfrac{19 (S^2 + 8)}{S (S^2 + 10)}$

Solutions to Comprehensive Exam
1. Answers to problem 1.

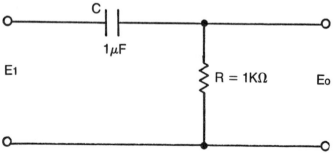

Fig. 9-25. Circuit for question 7 in Chapter 9.

a. High pass
b. Band pass
c. Band reject
d. High pass
e. Low pass
f. Low pass
g. All pass
h. All pass

2   a. $H(j\omega) = \dfrac{150\,j\omega}{-\omega^2 + 8\,j\omega + 12}$

b. $|H(j\omega)| = \dfrac{150\,\omega}{\sqrt{(12-\omega^2)^2 + (8\,\omega)^2}} = \dfrac{150\,\omega}{\sqrt{\omega^4 + 40\,\omega^2 + 144}}$

c. $\theta(\omega) = \arctan\dfrac{150\,\omega}{0} - \arctan\dfrac{8\,\omega}{-\omega^2 + 12}$

$\theta(\omega) = 90° - \arctan\dfrac{8\,\omega}{-\omega^2 + 12}$

3. $E_i = I_1\left(\dfrac{1}{SC}\right) + I_1 R - I_2 R = I_1\left(\dfrac{1}{SC}\right) I_1 + R - I_2 R$

$0 = -I_1 R + I_2(R + SL + R) = -I_1 R + I_2(2R + L)$

$\Delta = \begin{vmatrix} \dfrac{1}{SC} + R, & -R \\ -R, & 2R + SL \end{vmatrix} = \dfrac{2R}{SC} + \dfrac{L}{C} + R^2 + SLR$

$\Delta I_2 = \begin{vmatrix} \dfrac{1}{SC} + R, & E_i \\ -R, & 0 \end{vmatrix} = E_i R$

$E_o = I_2 R = \dfrac{\Delta I_2}{\Delta} R = \dfrac{E_i \times R \times R}{\Delta} = \dfrac{E_i R^2}{\Delta}$

$\dfrac{E_o}{E_i} = \dfrac{R^2}{\dfrac{2R}{SC} + \dfrac{L}{C} + SLR + R^2}$

$= \dfrac{S C R^2}{2R + SL + S^2 LCR + SCR^2}$

$= \dfrac{S C R^2}{S^2 LCR + S[L + CR^2] + 2R}$

4. Find $\theta\ (\omega)$

$$\theta\ (\omega)\ =\ \arctan \frac{0}{1} - \arctan \frac{2\omega}{1}$$

$$t_{(\omega)}\ =\ \frac{1}{1+\left(\frac{2\omega}{1}\right)^2}\ \frac{d}{d\omega}\ 2\omega$$

$$t_{(\omega)}\ =\ \frac{2}{1+4\omega^2}$$

$$t_{(3)}\ =\ \frac{2}{1+4\ (3)^2}$$

$$t_{(3)}\ =\ \frac{2}{37}\ \text{second.}$$

5. The circuit is a bridge-tee band-rejection filter.

$$RC\ =\ 6\times 10^{-3}\ \text{sec}$$

$$\frac{1}{RC}\ =\ 167$$

$$H_{(s)}\ =\ \frac{S^2 + \dfrac{2\ S}{RC} + \left(\dfrac{1}{RC}\right)^2}{S^2 + \dfrac{3S}{RC} + \left(\dfrac{1}{RC}\right)^2}\ =\ \frac{S + 334S + 27,000}{S^2 + 501S + 27,000}$$

$$\frac{E_o}{E_i}\ =\ \frac{\dfrac{R\ (j\,\omega L)}{R + j\,\omega\ L}}{R + j\,\omega\ L + \dfrac{R\ (j\,\omega L)}{R + j\omega L}}$$

$$\frac{E_o}{E_i}\ =\ \frac{R\ (j\,\omega L)}{R^2 + 2\ j\ \omega\ L\ R - \omega^2 L^2 + Rj\ \omega\ LR}\ \Bigg|\ \omega\ =\ \frac{R}{L}$$

$$=\ \frac{1}{3}$$

7. The scaling constant η is equal to:

$$n\ =\ \frac{\text{new cutoff}}{\text{old cutoff}}\ =\ \frac{5000}{1000}\ =\ 5$$

$$R_{new}\ =\ R_{old}\ =\ 1000\ \text{ohms.}$$

$$C_{new}\ =\ \frac{C_{old}}{\eta}\ =\ \frac{1\ \mu F}{5}\ =\ 0.2\ \mu F$$

8. A Butterworth filter magnitude falls off gradually, depending on the order of the filter. The response is a smooth decline in magnitude which falls at a rate of approximately n20dB/decade where **n** is the filter order.

A Chebychev low-pass filter has a number of lobes or variations in the magnitude response before the cutoff frequency. The number of lobes is dependent on the filter order.

9. Solution to problem 9a of exam

1. Take the function

$$Z_{(s)} = \frac{3S^4 + 9S^2 + 3}{S^3 + 2S}$$

and divide numerator by denominator:

$$
\begin{array}{r}
3S \phantom{{}+9S^2+3} \\
S^3 + 2S \,\overline{\big)\, 3S^4 + 9S^2 + 3} \\
\underline{3S^4 + 6S^2} \\
3S^2 + 3
\end{array}
$$

2. Write the impedance equation with 3S plus the remainder:

$$Z_{(s)} = 3S + \frac{3S^2 + 3}{S^3 + 2S} = 3S + \frac{3S^2 + 3}{S(S^2 + 2)}$$

3. Express the remainder in terms of partial fractions:

$$\frac{3S^2 + 3}{S(S^2 + 2)} = \frac{A}{S} + \frac{BS}{S^2 + 2}$$

$$3S^2 + 3 = A(S^2 + 2) + BS^2$$

4. Solve for B by letting S^2 become minus 2. Solve for A by letting S^2 equal 0.

$$
\begin{aligned}
S^2 &= -2 \\
3(-2) + 3 &= A(-2 + 2) + B(-2) \\
-6 + 3 &= 0 - 2B \\
B &= +1.5 \\
S^2 &= 0 \\
3(0) + 3 &= A(0 + 2) + B(0) \\
3 &= 2A \\
A &= 1.5
\end{aligned}
$$

5. Write the impedance equation including the partial fractions:

$$Z_{(s)} = 3S + \frac{1.5}{S} + \frac{1.5S}{S^2 + 2}$$

6. Comparing the impedance equation in step 5 to Foster's general equation we have

$$Z_{(s)} = K_a S + \frac{K_b}{S} + \frac{2K_c S}{S^2 + \omega_c^2}$$

Where

$K_a = 3$, $K_b = 1.5$, $2 K_c = 1.5$, $\omega_c^2 = 2$

7. Looking at the equation, we have the following: A capacitor of value equal to $\dfrac{1}{K_b}$ in series with an inductor of value K_a in series with a parallel circuit consisting of an inductor of value $\dfrac{2K_c}{\omega_c^2}$ in parallel with a capacitor of value $\dfrac{1}{2K_c}$.

8. By inspection of the equation in step 6, we find the following:

a. $L_1 = K_a = 3H$

b. $C_1 = \dfrac{1}{K_b} = \dfrac{1}{1.5} = 0.067\ F$

c. $L_2 = \dfrac{2K_c}{\omega_c^2} = \dfrac{1.5}{2} = 0.75H$

d. $C_2 = \dfrac{1}{2K_c} = \dfrac{1}{1.5} = .67F$

9. The complete circuit is shown in Fig. 9-26 Solution to problem 9b of exam.

1. Take the function

$$Z_{(s)} = \frac{S^4 + 12S^2 + 27}{S(S^2 + 6)}$$

and write the reciprocal:

$$Y_{(s)} = \frac{S(S^2 + 6)}{S^4 + 12S^2 + 27}$$

2. Factor the denominator of the fraction:

$$Y_{(s)} = \frac{S(S^2 + 6)}{(S^2 + 9)(S^2 + 3)}$$

3. Writing the equation $Y_{(s)}$ in factored form, we have

222

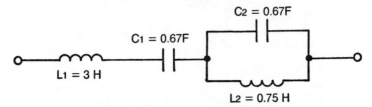

Fig. 9-26. Circuit for solution to 9A in Chapter 9.

$$Y_{(s)} = \frac{S(S^2 + 6)}{S^4 + 12S^2 + 27} = \frac{S(S^2 + 6)}{(S^2 + 9)(S^2 + 3)}$$

4. Expressing these in terms of partial fractions:

$$Y_{(s)} = \frac{S(S^2 + 6)}{(S^2 + 9)(S^2 + 3)} = \frac{AS}{S^2 + 9} + \frac{BS}{S^2 + 3}$$

5. Multipling both sides by $(S^2 + 9)(S^2 + 3)$ we have:

$$S(S^2 + 6) = A(S)(S^2 + 3) + B(S)(S^2 + 9)$$

$$S^2 + 6 = A(S^2 + 3) + B(S^2 + 9)$$

6. Let S^2 become -3, and solve for B:

$$-3 + 6 = A(-3 + 3) + B(-3 + 9)$$

$$3 = 0 + 6B$$

$$B = \frac{1}{2}$$

7. Let S^2 become -9, and solve for A:

$$-9 + 6 = A(-9 + 3) + B(-9 + 9)$$

$$-3 = -6A$$

$$A = \frac{1}{2}$$

8. Write the equation $Y_{(s)}$ in partial fractions:

$$Y_{(s)} = \frac{\frac{1}{2}S}{S^2 + 9} + \frac{\frac{1}{2}S}{S^2 + 3}$$

9. Compare the admittance equation in step 8 to Foster's general equation for method two, we have:

$$Y_{(s)} = K_a'S + \frac{K_b'}{S} + \frac{2K_c'S}{S^2 + \omega_c^2} + \frac{2K_d'S}{S^2 + \omega_d^2}$$

10. The result is shown in Fig. 9-27.

$$K_a' = 0, \ K_b' = 0, \ 2K_c' = \frac{1}{2}, \ 2K_d' = \frac{1}{2}$$

$$C_1 = \frac{2K_c'}{(\omega_c)^2} = \frac{\frac{1}{2}}{9} = \frac{1}{18} \text{ F} \qquad L_1 = \frac{1}{2K_c'} = \frac{1}{\frac{1}{2}} = 2H$$

$$C_2 = \frac{2K_d'}{(\omega_d)^2} = \frac{\frac{1}{2}}{3} = \frac{1}{6} \text{ F} \qquad L_2 = \frac{1}{2K_d'} = \frac{1}{\frac{1}{2}} = 2H$$

Solution to problem 9C.

$$Z_{(s)} = \frac{19 (S^2 + 8)}{S (S^2 + 10)}$$

Determine the partial fraction expansion,

$$\frac{19 (S^2 + 8)}{S (S^2 + 10)} = \frac{A}{S} = \frac{BS}{S^2 + 10}$$

Multiply both sides of equation by $S (S^2 + 10)$:

$$19 (S^2 + 8) = A (S^2 + 10) + B S^2$$

Let $S^2 = 0$,

$$19 (8) = A (10), A = 15.2$$

Let $S^2 = -10$,

$$19 (-10 + 8) = A (-10 + 10) + B (-10)$$

$$-38 = -10 B, B = 3.8$$

$$Z_{(s)} = \frac{15.2}{S} + \frac{3.8S}{S^2 + 10}$$

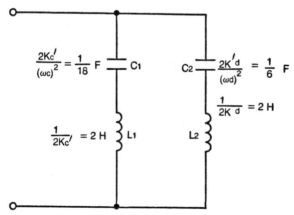

Fig. 9-27. Circuit for solution to 9B in Chapter 9.

We obtain the following:

$$K_a = 0, \ K_b = 15.2, \ 2K_c = 3.8, \ \omega_c^2 = 10$$

$$C_1 = \frac{1}{K_b} = \frac{1}{15.2} = 0.066f$$

$$C_2 = \frac{1}{2K_c} = \frac{1}{3.8} = 0.263F$$

$$L_2 = \frac{2K_c}{\omega_c^2} = \frac{3.8}{10} = 0.38 \ H$$

The circuit is shown in Fig. 9-28.
Solution to problem 10a.

1. Take the function $\quad Z_{(s)} = \dfrac{3S^4 + 9S^2 + 3}{S^3 + 2S}$

and write the reciprocal:

$$\frac{1}{Z_{(s)}} = Y_{(s)} = \frac{S^3 + 2S}{3S^4 + 9S^2 + 3}$$

2. It is obvious that the function denominator cannot be factored without use of the quadratic formula. Therefore, using the quadratic formula, factor the denominator into a function of the form $(S + q)(S + r)$.

The quadratic formula is:

$$\chi = \frac{-b \pm \sqrt{b^2 - 4ac}}{2a}$$

Where χ is the root of the equation $a\chi^2 + b\chi + c = 0$.

$$a = 3, b = 9, c = 3$$
$$\chi = \frac{-9 \pm \sqrt{(9)^2 - 4(3)(3)}}{2(3)}$$
$$= -1.5 \pm 1.15$$

The roots work out to be -2.65 and -0.35.

3. Writing the equation Y_s in factored form, we have

$$Y_{(s)} = \frac{S^3 + 2S}{3S^4 + 9S^2 + 3}$$

$$= \frac{S^3 + 2S}{3(S^2 + .35)(S^2 + 2.65)}$$

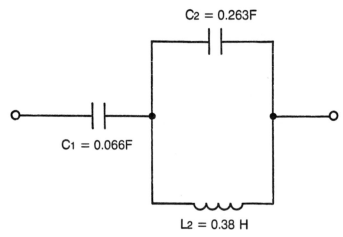

Fig. 9-28. Circuit for solution to 9C in Chapter 9.

4. Expressing these in terms of partial fractions:

$$Y_{(s)} = \frac{AS}{S^2 + 2.65} + \frac{BS}{S^2 + 0.35}$$

5. Multiplying each side of the equation by $3(S^2 + .35)(S^2 + 2.65)$, we have

$$S^3 + 2S = 3AS(S^2 + 0.35) + 3BS(S^2 + 2.65)$$

or,

$$S^2 + 2 = 3A(S^2 + 0.35) + 3B(S^2 + 2.65)$$

6. Let S^2 become -0.35 to determine B.

$$-0.35 + 2 = 3A(-0.35 + 0.35) + 3B(-0.35 + 2.65)$$
$$1.65 = 0 + (2.30)(3)B$$
$$B = 0.239$$

7. Let S^2 become -2.65 to determine A.

$$-2.65 + 2 = 3A(-2.65 + .35) + 3B(-2.65 + 2.65)$$
$$-.65 = 3A(-2.3) + 0$$
$$A = 0.094$$

8. Write the equation $Y_{(s)}$ in terms of partial fractions:

$$Y_{(s)} = \frac{0.094\,S}{S^2 + 2.65} + \frac{0.239\,S}{S^2 + 0.35}$$

9. Comparing the admittance equation in step 8 to Foster's general equation for method 2, we have

$$Y(S) = K_a' S + \frac{K_b'}{S} + \frac{2 K_c' S}{S^2 + \omega_c^2} + \frac{2 K_d' S}{S^2 + \omega_d^2}$$

where $K_a' = 0$, $K_b' = 0$, $2 K_c' = 0.094$, $2 K_d' = 0.239$.

10. Looking at the equation we have the following: a capacitor of value $\dfrac{2K_c'}{\omega_c^2}$ in series with an inductor of value $\dfrac{1}{2K_c'}$, all in parallel with a capacitor of value $\dfrac{2K_d'}{\omega_d^2}$ in series with an inductor of value $\dfrac{1}{2K_d'}$.

11. By inspection of the equation in step 9, we have the following:

a. $L_1 = \dfrac{1}{2K_c'} = \dfrac{1}{0.094} = 10.62 \text{ H}$

b. $C_1 = \dfrac{2K_c'}{\omega_c^2} = \dfrac{0.094}{2.65} = 0.0355 \text{F}$

c. $L_2 = \dfrac{1}{2K_d'} = \dfrac{1}{0.239} = 4.18 \text{ H}$

d. $C_2 = \dfrac{2K_d'}{\omega_d^2} = \dfrac{.239}{0.35} = 0.684 \text{F}$

12. The complete circuit is shown in Fig. 9-29.

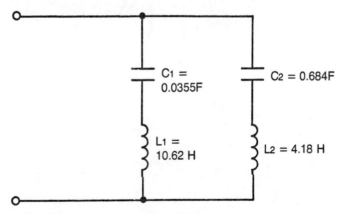

Fig. 9-29. Circuit for solution to 10A in Chapter 9.

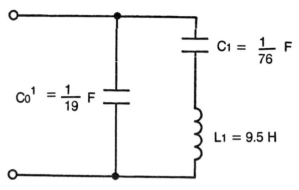

Fig. 9-30. Circuit for solution to 10B in Chapter 9.

Solution to Problem 10 B of exam.

$$Z_{(S)} = \frac{19(S^2 + 8)}{S(S^2 + 10)}$$

Taking the reciprocal to get $Y_{(S)}$, we have

$$Y_{(S)} = \frac{S(S^2 + 10)}{19(S^2 + 8)} = \frac{S^3 + 10S}{19(S^2 + 8)}$$

$$
\begin{array}{r}
\frac{S}{19} \\
19(S^2 + 8) \overline{\smash{\big)}\ S^3 + 10S} \\
\underline{S^3 + 8S} \\
2S
\end{array}
$$

$$Y_{(S)} = \frac{S}{19} + \frac{2S}{19(S^2 + 8)} = \left(\frac{1}{19}\right) S + \frac{\frac{2}{19}S}{S^2 + 8}$$

where $K_a' = \frac{1}{19}$, $K_b' = 0$, $2K_c' = \frac{2}{19}$, $\omega_c^2 = 8$

$$L_1 = \frac{1}{2K_c'} = 9.5 \text{ H}$$

$$C_1 = \frac{2K_c'}{\omega_c^2} = \frac{2}{152} = \frac{1}{76} \text{ F}$$

$$C_a' = K_a' = \frac{1}{19} \text{ F}$$

circuit is shown in Fig. 9-30.

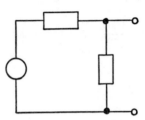

Index